5

Module 4

Eureka Math™
A Story of Units

Camila Gutierrez

Special thanks go to the Gordan A. Cain Center and to the Department of Mathematics at Louisiana State University for their support in the development of *Eureka Math*.

Published by Common Core

Copyright © 2014 Common Core, Inc. All rights reserved. No part of this work may be reproduced or used in any form or by any means – graphic, electronic, or mechanical, including photocopying or information storage and retrieval systems – without written permission from the copyright holder. "Common Core" and "Common Core, Inc.," are registered trademarks of Common Core, Inc.

Common Core, Inc. is not affiliated with the Common Core State Standards Initiative.

Printed in the U.S.A.

This book may be purchased from the publisher at commoncore.org

10 9 8 7 6 5 4 3 2 1

ISBN 978-1-63255-036-1

Name _____   Date _____

1. Estimate the length of your pencil to the nearest inch. _____

2. Using a ruler, measure your pencil strip to the nearest $\frac{1}{2}$ inch and mark the measurement with an X above the ruler below. Construct a line plot of your classmates' pencil measurements.

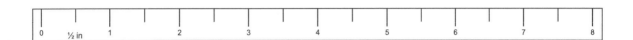

3. Using a ruler, measure your pencil strip to the nearest $\frac{1}{4}$ inch and mark the measurement with an X above the ruler below. Construct a line plot of your classmates' pencil measurements.

4. Using a ruler, measure your pencil strip to the nearest $\frac{1}{8}$ inch and mark the measurement with an X above the ruler below. Construct a line plot of your classmates' pencil measurements.

EUREKA MATH™

Lesson 1:   Measure and compare pencil lengths to the nearest $\frac{1}{2}$, $\frac{1}{4}$, and $\frac{1}{8}$ of an inch, and analyze the data through line plots.

1

5. Use all three of your line plots to complete the following.

   a. Compare the three plots, and write one sentence that describes how the plots are alike and one sentence that describes how they are different.

   b. What is the difference between the measurements of the longest and shortest pencils on each of the three line plots?

   c. Write a sentence describing how you could create a more precise ruler to measure your pencil strip.

EUREKA
MATH™

Lesson 1: Measure and compare pencil lengths to the nearest $\frac{1}{2}$, $\frac{1}{4}$, and $\frac{1}{8}$ of an inch, and analyze the data through line plots.

2

Name _____ Date _____

A meteorologist set up rain gauges at various locations around a city and recorded the rainfall amounts in the table below. Use the data in the table to create a line plot using $\frac{1}{8}$ inches.

Location	Rainfall Amount (inches)
1	$\frac{1}{8}$
2	$\frac{3}{8}$
3	$\frac{3}{4}$
4	$\frac{3}{4}$
5	$\frac{1}{4}$
6	$1\frac{1}{4}$
7	$\frac{1}{8}$
8	$\frac{1}{4}$
9	1
10	$\frac{1}{8}$

a. Which location received the most rainfall?

b. Which location received the least rainfall?

c. Which rainfall measurement was the most frequent?

d. What is the total rainfall in inches?

EUREKA
MATH™

Lesson 1: Measure and compare pencil lengths to the nearest $\frac{1}{2}$, $\frac{1}{4}$, and $\frac{1}{8}$ of an inch, and analyze the data through line plots.

Name _____   Date _____

1. Draw a picture to show the division. Write a division expression using unit form. Then, express your answer as a fraction. The first one is partially done for you.

    a. $1 \div 5 = 5$ fifths $\div 5 = 1$ fifth $= \frac{1}{5}$

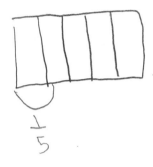

    b. $3 \div 4 = 12$ fourths $\div 4 = 3$ fourths $= \frac{3}{4}$

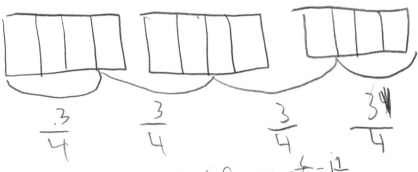

$$\frac{3}{4} \qquad \frac{3}{4} \qquad \frac{3}{4} \qquad \frac{3}{4}$$

    c. $6 \div 4 = 24$ fourths $\div 4 = 6$ fourths $= \frac{6}{4} = 1\frac{1}{2}$

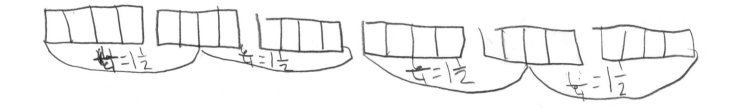

$\frac{6}{4} = 1\frac{1}{2} \qquad \frac{6}{4} = 1\frac{1}{2} \qquad \frac{6}{4} = 1\frac{1}{2} \qquad \frac{6}{4} = 1\frac{1}{2}$

2. Draw to show how 2 children can equally share 3 cookies. Write an equation, and express your answer as a fraction.

$3 \div 2 = 6$ halves $\div 2 = 3$ halves $= \frac{3}{2} = 1\frac{1}{2}$

$\frac{3}{2} = 1\frac{1}{2}$

Each child will recieve $1\frac{1}{2}$ cookies

3. Carly and Gina read the following problem in their math class.

    Seven cereal bars were shared equally by 3 children. How much did each child receive?

    Carly and Gina solve the problem differently. Carly gives each child 2 whole cereal bars, and then divides the remaining cereal bar between the 3 children. Gina divides all the cereal bars into thirds and shares the thirds equally among the 3 children.

    a. Illustrate both girls' solutions.

    b. Explain why they are both right.

4. Fill in the blanks to make true number sentences.

a. $2 \div 3 = \dfrac{2}{3}$

b. $15 \div 8 = \dfrac{15}{8}$

$\dfrac{15 - 8}{8} = \dfrac{15}{8}$

c. $11 \div 4 = \dfrac{11}{4}$

d. $\dfrac{3}{2} = \underline{3} \div \underline{2}$

e. $\dfrac{9}{13} = \underline{9} \div \underline{13}$

f. $1\dfrac{1}{3} = \underline{4} \div \underline{3}$

Name _____ Date _____

1. Draw a picture to show the division. Express your answer as a fraction.

a. $1 \div 4 = 4$ fourths $\div 4 = 1$ fourth $= \frac{1}{4}$

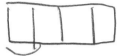

$\frac{1}{4}$

b. $3 \div 5 = 15$ fifths $\div 5 = 3$ fifths $= \frac{3}{5}$

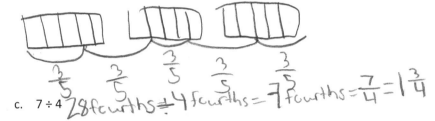

$\frac{3}{5}$ $\frac{3}{5}$ $\frac{3}{5}$ $\frac{3}{5}$ $\frac{3}{5}$

c. $7 \div 4 = 28$ fourths $\div 4 = 7$ fourths $= \frac{7}{4} = 1\frac{3}{4}$

$\frac{7}{4} = 1\frac{3}{4}$ $\frac{7}{4} = 1\frac{3}{4}$ $\frac{7}{4} = 1\frac{3}{4}$ $\frac{7}{4} = 1\frac{3}{4}$

2. Using a picture, show how six people could share four sandwiches. Then, write an equation and solve.

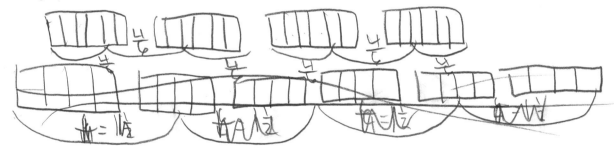

$6 \div 4 = 24$ fourths $\div 4 = 6$ fourths $= \frac{6}{4} = 1\frac{1}{2}$

$4 \div 6 = 24$ sixths $\div 6 = 4$ sixths $= \frac{4}{6} = \frac{2}{3}$

Everyone gets $\frac{2}{3}$ sandwiches

3. Fill in the blanks to make true number sentences.

a. $2 \div 7 = \dfrac{2}{7}$

b. $39 \div 5 = \dfrac{39}{5}$

c. $13 \div 3 = \dfrac{13}{3}$

d. $\dfrac{9}{5} = 9 \div 5$

e. $\dfrac{19}{28} = 19 \div 28$

f. $1\dfrac{3}{5} = 8 \div 5$

Name _____   Date _____

1. Fill in the chart. The first one is done for you.

Division Expression	Unit Forms	Improper Fraction	Mixed Numbers	Standard Algorithm (Write your answer in whole numbers and fractional units. Then check.)
a. $5 \div 4$	20 fourths ÷ 4 = 5 fourths	$\frac{5}{4}$	$1\frac{1}{4}$	$4\overline{\smash{\big)}\,5}$ with $1\frac{1}{4}$ on top, -4, remainder 1 Check $4 \times 1\frac{1}{4} = 1\frac{1}{4} + 1\frac{1}{4} + 1\frac{1}{4} + 1\frac{1}{4}$ $= 4 + \frac{4}{4}$ $= 4 + 1$ $= 5$
b. $3 \div 2$	6 halves ÷ 2 = 3 halves	$\frac{3}{2}$	$1\frac{1}{2}$	$2\overline{\smash{\big)}\,3}$ with $1\frac{1}{2}$ on top, 2, remainder 1 Check $2 \times 1\frac{1}{2} = 1\frac{1}{2} + 1\frac{1}{2}$ $= 4\frac{4}{2} \ 2 + \frac{2}{2}$ $= 4 + \frac{2}{2} \ 2 + 1$ $= 6 = 3$
c. $6 \div 4$	24 fourths ÷ 4 = 6 fourths	$\frac{6}{4}$	$1\frac{1}{2}$	$4\overline{\smash{\big)}\,6}$ with $1\frac{2}{4} = 1\frac{1}{2}$ on top, 4, remainder 2 Check $4 \times 1\frac{1}{2} =$ $= 1\frac{1}{2} + 1\frac{1}{2} + 1\frac{1}{2} + 1\frac{1}{2}$ $= 4 + \frac{4}{2} = 2$ $= 4 + 2$ $=$
d. $5 \div 2$	10 halves ÷ 2 = 5 halves	$\frac{5}{2}$	$2\frac{1}{2}$	$2\overline{\smash{\big)}\,5}$ with $2\frac{1}{2}$ on top, 4, remainder 1 Check $2 \times 2\frac{1}{2}$ $= 2\frac{1}{2} + 2\frac{1}{2}$ $= 4 + \frac{2}{2}$ $= 5$

2. A principal evenly distributes 6 reams of copy paper to 8 fifth-grade teachers.
    a. How many reams of paper does each fifth-grade teacher receive? Explain how you know using pictures, words, or numbers.

$6 \div 8 = \frac{6}{8} = \frac{3}{4}$ Each 5-grade teacher recieves $\frac{3}{4}$ reams of copy papaper.

    b. If there were twice as many reams of paper and half as many teachers, how would the amount each teacher receives change? Explain how you know using pictures, words, or numbers.

$12 \div 4 = 3$ Each teacher recieves 3 reams of copy paper

3. A caterer has prepared 16 trays of hot food for an event. The trays are placed in warming boxes for delivery. Each box can hold 5 trays of food.

    a. How many warming boxes are necessary for delivery if the caterer wants to use as few boxes as possible? Explain how you know.

$16 \div 5 = \frac{16}{5} = 3\frac{1}{5}$

The caterer needs four boxes.

    b. If the caterer fills a box completely before filling the next box, what fraction of the last box will be empty?

$\frac{4}{5}$ of the box is empty.

Name _____   Date _____

1. Fill in the chart. The first one is done for you.

Division Expression	Unit Forms	Improper Fractions	Mixed Numbers	Standard Algorithm (Write your answer in whole numbers and fractional units. Then check.)
a. $4 \div 3$	12 thirds ÷ 3 = 4 thirds	$\dfrac{4}{3}$	$1\dfrac{1}{3}$	$3 \overline{)\,4\,}$ with $1\frac{1}{3}$ above, -3, remainder 1 Check $3 \times 1\frac{1}{3} = 1\frac{1}{3} + 1\frac{1}{3} + 1\frac{1}{3}$ $= 3 + \frac{3}{3}$ $= 3 + 1$ $= 4$
b. $7 \div 5$	~~35~~ fifths ÷ 5 = 7 fifths	$\dfrac{7}{5}$	$1\dfrac{2}{5}$	$5\overline{)\,7\,}$ with $1\frac{2}{5}$ above, -5, remainder 2 Check $5 \times 1\frac{2}{5}$ $1\frac{2}{5}+1\frac{2}{5}+1\frac{2}{5}+1\frac{2}{5}+1\frac{2}{5}$ $= 5\frac{10}{5}$ $= 7$
c. $7 \div 2$	~~14~~ halves ÷ 2 = 7 halves	$\dfrac{7}{2}$	$3\dfrac{1}{2}$	$2\overline{)\,7\,}$ with $3\frac{1}{2}$ above, 6, remainder 1 Check $2 \times 3\frac{1}{2}$ $= 3\frac{1}{2}+3\frac{1}{2}$ $= 6 + \frac{2}{2}$ $= 7$
d. $7 \div 4$	~~28~~ fourths ÷ 4 = 7 fourths	$\dfrac{7}{4}$	$1\dfrac{3}{4}$	$4\overline{)\,7\,}$ with $1\frac{3}{4}$ above, 4, remainder 3 Check $4 \times 1\frac{3}{4}$ $= 1\frac{3}{4}+1\frac{3}{4}+1\frac{3}{4}+1\frac{3}{4}$ $= 4 + \frac{12}{4}$ $= 7$

2. A coffee shop uses 4 liters of milk every day.

   a. If there are 15 liters of milk in the refrigerator, after how many days will more milk need to be purchased? Explain how you know.

   $$15 \div 4 = \frac{15}{4} = 3\frac{3}{4}$$

   In 3 days More milk will need to be purchased.

   b. If only half as much milk is used each day, after how many days will more milk need to be purchased?

   $$15 \div 2 = \frac{15}{2} = 7\frac{1}{2}$$

   In 7½ more milk will need to be purchased.

3. Polly buys 14 cupcakes for a party. The bakery puts them into boxes that hold 4 cupcakes each.

   a. How many boxes will be needed for Polly to bring all the cupcakes to the party? Explain how you know.

   $$14 \div 4 = \frac{14}{4} = 3\frac{2}{4} = 3\frac{1}{2}$$

   Polly will need 4 boxes.

   b. If the bakery completely fills as many boxes as possible, what fraction of the last box is empty? How many more cupcakes are needed to fill this box?

   $\frac{1}{2}$ of the box ~~need~~ is empty.
   2 more cupcakes are full

Name _____   Date _____

1. Draw a tape diagram to solve. Express your answer as a fraction. Show the multiplication sentence to check your answer. The first one is done for you.

a. $1 \div 3 = \frac{1}{3}$

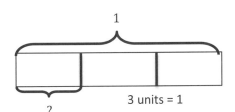

3 units = 1

1 unit = 1 ÷ 3

$= \frac{1}{3}$

Check: $3 \times \frac{1}{3}$

$= \frac{1}{3} + \frac{1}{3} + \frac{1}{3}$

$= \frac{3}{3}$

$= 1$

b. $2 \div 3 = \underline{\quad}$

c. $7 \div 5 = \underline{\quad}$

d. $14 \div 5 = \underline{\quad}$

2. Fill in the chart. The first one is done for you.

Division Expression	Fraction	Between which two whole numbers is your answer?	Standard Algorithm
a. $13 \div 3$	$\dfrac{13}{3}$	4 and 5	$3\overline{)13}$ with quotient $4\frac{1}{3}$, -12, remainder 1
b. $6 \div 7$		0 and 1	$7\overline{)6}$
c. ___ ÷ ___	$\dfrac{55}{10}$		
d. ___ ÷ ___	$\dfrac{32}{40}$		$40\overline{)32}$

3. Greg spent $4 on 5 packs of sport cards.
    a. How much did Greg spend on each pack?

    b. If Greg spent half as much money and bought twice as many packs of cards, how much did he spend on each pack? Explain your thinking.

4. Five pounds of birdseed is used to fill 4 identical bird feeders.
    a. What fraction of the birdseed will be needed to fill each feeder?

    b. How many pounds of birdseed are used to fill each feeder? Draw a tape diagram to show your thinking.

    c. How many ounces of birdseed are used to fill three birdfeeders?

Name _____   Date _____

1. Draw a tape diagram to solve. Express your answer as a fraction. Show the addition sentence to support your answer. The first one is done for you.

a. $1 \div 4 = \frac{1}{4}$

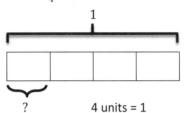

?   4 units = 1

1 unit = $1 \div 4$

$= \frac{1}{4}$

Check:

$\begin{array}{r} 0 \quad \frac{1}{4} \\ 4 \overline{\smash{\big)}\ 1} \\ -\ 0 \\ \hline 1 \end{array}$

$4 \times \frac{1}{4}$

$= \frac{1}{4} + \frac{1}{4} + \frac{1}{4} + \frac{1}{4}$

$= \frac{4}{4}$

$= 1$

b. $4 \div 5 = \overline{}$

c. $8 \div 5 = \overline{}$

d. $14 \div 3 = \overline{}$

2. Fill in the chart. The first one is done for you.

Division Expression	Fraction	Between which two whole numbers is your answer?	Standard Algorithm
a. $16 \div 5$	$\dfrac{16}{5}$	3 and 4	$3\ \tfrac{1}{5}$ $5\,\overline{)\ 16}$ $\quad -15$ $\qquad 1$
b. ____ ÷ ____	$\dfrac{3}{4}$	0 and 1	⌐
c. ____ ÷ ____	$\dfrac{7}{2}$		$2\,\overline{)\ 7}$
d. ____ ÷ ____	$\dfrac{81}{90}$		⌐

EUREKA MATH   Lesson 4:   Use tape diagrams to model fractions as division.

17

3. Jackie cut a 2-yard spool into 5 equal lengths of ribbon.
    a. What is the length of each ribbon in yards? Draw a tape diagram to show your thinking.

    b. What is the length of each ribbon in feet? Draw a tape diagram to show your thinking.

4. Baa Baa, the black sheep, had 7 pounds of wool. If he separated the wool equally into 3 bags, how much wool would be in 2 bags?

5. An adult sweater is made from 2 pounds of wool. This is 3 times as much wool as it takes to make a baby sweater. How much wool does it take to make a baby sweater? Use a tape diagram to solve.

Name _____ Date _____

1. A total of 2 yards of fabric is used to make 5 identical pillows. How much fabric is used for each pillow?

2 : 5

$\frac{2}{5}$

Each pillow will use $\frac{2}{5}$ of a yard.

2. An ice-cream shop uses 4 pints of ice cream to make 6 sundaes. How many pints of ice cream are used for each sundae?

4 : 6

$\frac{4}{6} = \frac{2}{3}$

$\frac{2}{3}$ pints of ice cream are used for each sundae.

3. An ice-cream shop uses 6 bananas to make 4 identical sundaes. How many bananas are used in each sundae? Use a tape diagram to show your work.

4 ÷ 4

$\frac{6}{4} = 1\frac{1}{2}$

$1\frac{1}{2}$ bananas are used for each sundae

EUREKA MATH | Lesson 5: Solve word problems involving the division of whole numbers with answers in the form of fractions or whole numbers.

19

4. Julian has to read 4 articles for school. He has 8 nights to read them. He decides to read the same number of articles each night.

   a. How many articles will he have to read per night?

   $4 \div 8$

   $\dfrac{4}{8} \div \dfrac{1}{2}$

   Julian has to read $\frac{1}{2}$ of an article each night

   b. What fraction of the reading assignment will he read each night?

   $\dfrac{1}{8}$

   He will read $\frac{1}{8}$ of the assignment each night

5. 40 students shared 5 pizzas equally. How much pizza will each student receive? What fraction of the pizza did each student receive?

   An $5 \div 40$

   $\dfrac{5}{40} = \dfrac{1}{8}$

   Each student receives $\frac{1}{8}$ of a pizza.

   4 liters

6. Lillian had 2 two-liter bottles of soda, which she distributed equally between 10 glasses.

   a. How much soda was in each glass? Express your answer as a fraction of a liter.

   $4 \div 10$

   $\dfrac{4 \div 2}{10 \div 2} = \dfrac{2}{5}$ liter

   Each glass holds $\frac{2}{5}$ of liter.

EUREKA MATH

Lesson 5:   Solve word problems involving the division of whole numbers with answers in the form of fractions or whole numbers.

20

b.   Express your answer as a decimal number of liters.

$$0.4 \text{ liters}$$

1000

c.   Express your answer as a whole number of milliliters.

$$14 \div 3$$
$$\frac{14}{3} = 4\frac{2}{3}$$

7.   The Calef family likes to paddle along the Susquehanna River.

a.   They paddled the same distance each day over the course of 3 days, traveling a total of 14 miles. How many miles did they travel each day? Show your thinking in a tape diagram.

They traveled $4\frac{2}{3}$ miles each day.

$2\frac{1}{3}$

b.   If the Calefs went half their daily distance each day, but extended their trip to twice as many days, how far would they travel?

$6 \times 2\frac{1}{3}$

$2\frac{1}{3} + 2\frac{1}{3} + 2\frac{1}{3} + 2\frac{1}{3} + 2\frac{1}{3} + 2\frac{1}{3}$

$= 12 + 2$

$= 14$

The Calefs would travel 14 miles

Name _____   Date _____

1. When someone donated 14 gallons of paint to Rosendale Elementary School, the fifth-grade decided to
    use it to paint murals. They split the gallons equally among the four classes.
    a. How much paint did each class have to paint their mural?

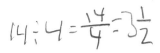

$$14 \div 4 = \frac{14}{4} = 3\frac{1}{2}$$

Each class
has $3\frac{1}{2}$ gallons
of paint

    b. How much paint will three classes use? Show your thinking using words, numbers, or pictures.

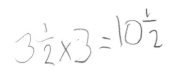

$$3\frac{1}{2} \times 3 = 10\frac{1}{2}$$

3 classes will
use $10\frac{1}{2}$ gallons
of paint.

    c. If 4 students share a 30 square foot wall equally, how many square feet of the wall will be painted by
        each student?

$$30 \div 4 = \frac{30}{4} = 7\frac{2}{4} = 7\frac{1}{2}$$

Each student
will paint
$7\frac{1}{2}$ sq ft

    d. What fraction of the wall will each student paint?

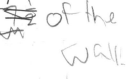

of the
wall

EUREKA
MATH   Lesson 5:   Solve word problems involving the division of whole numbers with
                            answers in the form of fractions or whole numbers.

                                                                                              22

2. Craig bought a 3-foot long baguette, and then made 4 equally sized sandwiches with it.
   a. What portion of the baguette was used for each sandwich? Draw a visual model to help you solve this problem.

   $3 \div 4 = \frac{3}{4}$ $\frac{3}{4}$ of the baguette was used for each sandwich.

   b. How long, in feet, is one of Craig's sandwiches?

   $\frac{3 \times 3}{4 \times 3} = \frac{9}{12} = \frac{1}{4}$ $\frac{1}{12}$ of a foot.

   c. How many inches long is one of Craig's sandwiches?

   $\overset{are 3}{\underset{\wedge}{Craig's}}$ 9 inches sandwiches

3. Scott has 6 days to save enough money for a $45 concert ticket. If he saves the same amount each day, what is the minimum amount he must save each day in order to reach his goal? Express your answer in dollars.

   $45 \div 6 = \frac{45}{6} = 7\frac{1}{2}$

   Scott must at least $7.50 every day to reach his goal

Name _____   Date _____

1. Find the value of each of the following.

a.

△ | △ | △
△ | △ | △
△ | △ | △

$\frac{1}{3}$ of 9 =

$\frac{2}{3}$ of 9 =

$\frac{3}{3}$ of 9 =

b.

△ △ △ △
△ △ △ △
△ △ △ △

$\frac{1}{3}$ of 15 =

$\frac{2}{3}$ of 15 =

$\frac{3}{3}$ of 15 =

c.

△ △ | △ △ | △ △ | △ △ | △ △
△ △ | △ △ | △ △ | △ △ | △ △

$\frac{1}{5}$ of 20=

$\frac{4}{5}$ of 20 =

$\frac{}{5}$ of 20 = 20

d.

△ △ | △ △ | △ △ | △ △ | △ △ | △ △ | △ △ | △ △
△ | △ | △ | △ | △ | △ | △ | △

$\frac{1}{8}$ of 24 = $\frac{6}{8}$ of 24 =

$\frac{3}{8}$ of 24 = $\frac{7}{8}$ of 24 =

$\frac{4}{8}$ of 24 =

EUREKA
MATH™

Lesson 6:   Relate fractions as division to fraction of a set.

25

2. Find $\frac{4}{7}$ of 14. Draw a set and shade to show your thinking.

3. How does knowing $\frac{1}{8}$ of 24 help you find three-eighths of 24? Draw a picture to explain your thinking.

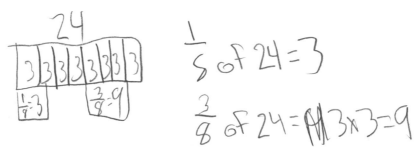

$\frac{1}{8}$ of 24 = 3

$\frac{3}{8}$ of 24 = 3×3 = 9

4. There are 32 students in a class. Of the class, $\frac{3}{8}$ of the students bring their own lunches. How many students bring their lunches?

12 students

5. Jack collected 18 ten dollar bills while selling tickets for a show. He gave $\frac{1}{6}$ of the bills to the theater and kept the rest. How much money did he keep?

$180

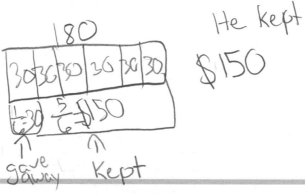

He kept
$150

EUREKA MATH

Lesson 6:   Relate fractions as division to fraction of a set.

26

Name _____   Date _____

1.   Find the value of each of the following.

a.

$\frac{1}{3}$ of 12 =

$\frac{2}{3}$ of 12 =

$\frac{3}{3}$ of 12 =

b.

$\frac{1}{4}$ of 20 = $\frac{3}{4}$ of 20 =

$\frac{2}{4}$ of 20 = $\frac{4}{4}$ of 20 =

c.

$\frac{1}{5}$ of 35 = $\frac{3}{5}$ of 35 = $\frac{5}{5}$ of 35 =

$\frac{2}{5}$ of 35 = $\frac{4}{5}$ of 35 = $\frac{6}{5}$ of 35 =

$24 \times 6 =$ [handwritten]

2. Find $\frac{2}{3}$ of 18. Draw a set and shade to show your thinking.

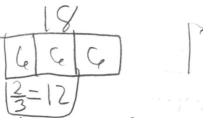

3. How does knowing $\frac{1}{5}$ of 10 help you find $\frac{3}{5}$ of 10? Draw a picture to explain your thinking.

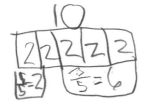

$\frac{1}{5}$ of $10 = 2$

$\frac{3}{5}$ of $10 = 2 \times 3 = 6$

4. Sara just turned 18 years old. She spent $\frac{4}{9}$ of her life living in Rochester, NY. How many years did Sara live in Rochester?

Sara lived 8 years in Rochester, NY

5. A farmer collected 12 dozen eggs from her chickens. She sold $\frac{5}{6}$ of the eggs at the farmers' market, and gave the rest to friends and neighbors.

a. How many dozens did the farmer give away? How many eggs did she give away?

10 dozens,
120 eggs.

b. She sold each dozen for $4.50. How much did she earn from the eggs she sold?

$10 \times \$4.50 =$

She earned $45

Name _____   Date _____

1. Solve using a tape diagram.

a. $\frac{1}{3}$ of 18

b. $\frac{1}{3}$ of 36

c. $\frac{3}{4} \times 24$

d. $\frac{3}{8} \times 24$

e. $\frac{4}{5} \times 25$

f. $\frac{1}{7} \times 140$

g. $\frac{1}{4} \times 9$

h. $\frac{2}{5} \times 12 =$

i. $\frac{2}{3}$ of a number is 10. What's the number?

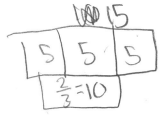

j. $\frac{3}{4}$ of a number is 24. What's the number?

EUREKA
MATH™

Lesson 7:   Multiply any whole number by a fraction using tape diagrams.

29

2. Solve using tape diagrams.

a. There are 48 students going on a field trip. One-fourth are girls. How many boys are going on the trip?

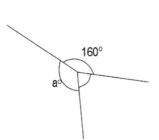

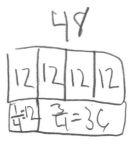

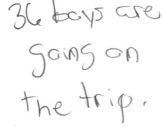

36 boys are going on the trip.

b. Three angles are labeled below with arcs. The smallest angle is $\frac{3}{8}$ as large as the 160° angle. Find the value of angle a.

c. Abbie spent $\frac{5}{8}$ of her money and saved the rest. If she spent $45, how much money did she have at first?

Abbie had $72 at first.

d. Mrs. Harrison used 16 ounces of dark chocolate while baking. She used $\frac{2}{5}$ of the chocolate to make some frosting and used the rest to make brownies. How much more chocolate did Mrs. Harrison use in the brownies than in the frosting?

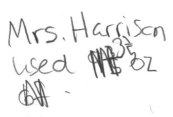

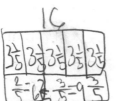

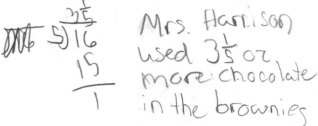

Mrs. Harrison used $3\frac{1}{5}$ oz more chocolate in the brownies

EUREKA MATH

Lesson 7: Multiply any whole number by a fraction using tape diagrams.

30

Name _____   Date _____

1. Solve using a tape diagram.

    a. $\frac{1}{4}$ of 24

    b. $\frac{1}{4}$ of 48

    c. $\frac{2}{3} \times 18$

    d. $\frac{2}{6} \times 18$

    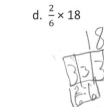

    e. $\frac{3}{7} \times 49$

    f. $\frac{3}{10} \times 120$

    g. $\frac{1}{3} \times 31$

    h. $\frac{2}{5} \times 20$

    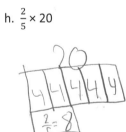 

    i. $\frac{1}{4} \times 25$

    j. $\frac{3}{4} \times 25$

    k. $\frac{3}{4}$ of a number is 27. What's the number?

    l. $\frac{2}{5}$ of a number is 14. What's the number?

    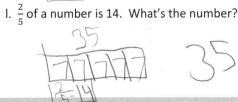

EUREKA MATH

Lesson 7:   Multiply any whole number by a fraction using tape diagrams.

31

2. Solve using tape diagrams.

   a. A skating rink sold 66 tickets. Of these, $\frac{2}{3}$ were children's tickets, and the rest were adult tickets. What total number of adult tickets were sold?

   b. A straight angle is split into two smaller angles as shown. The smaller angle's measure is $\frac{1}{6}$ that of a straight angle. What is the value of angle a?

   c. Annabel and Eric made 17 ounces of pizza dough. They used $\frac{5}{8}$ of the dough to make a pizza and used the rest to make calzones. What is the difference between the amount of dough they used to make pizza, and the amount of dough they used to make calzones?

   d. The New York Rangers hockey team won $\frac{3}{4}$ of their games last season. If they lost 21 games, how many games did they play in the entire season?

EUREKA MATH™   Lesson 7:   Multiply any whole number by a fraction using tape diagrams.

32

Name _____   Date _____

1. Laura and Sean find the product of $\frac{2}{3} \times 4$ using different methods.

    Laura: It's 2 thirds of 4.   *Sean:* It's 4 groups of 2 thirds.

    $$\frac{2}{3} \times 4 = \frac{4}{3} + \frac{4}{3} = 2 \times \frac{4}{3} = = \frac{8}{3}$$ $$\frac{2}{3} + \frac{2}{3} + \frac{2}{3} + \frac{2}{3} = 4 \times \frac{2}{3} = \frac{8}{3}$$

    Use words, pictures, or numbers to compare their methods in the space below.

2. Rewrite the following addition expressions as fractions as shown in the example.

    Example: $\frac{2}{3} + \frac{2}{3} + \frac{2}{3} + \frac{2}{3} = \frac{4 \times 2}{3} = \frac{8}{3}$

    a. $\frac{7}{4} + \frac{7}{4} + \frac{7}{4} =$ b. $\frac{14}{5} + \frac{14}{5} =$ c. $\frac{4}{7} + \frac{4}{7} + \frac{4}{7} =$

3. Solve and model each problem as a fraction of a set and as repeated addition.

    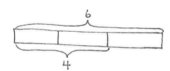   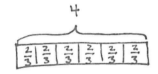

    Example: $\frac{2}{3} \times 6 = 2 \times \frac{6}{3} = 2 \times 2 = 4.$ $6 \times \frac{2}{3} = \frac{6 \times 2}{3} = 4$

    a. $\frac{1}{2} \times 8$ $8 \times \frac{1}{2}$

    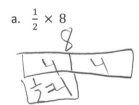   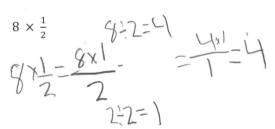

    b. $\frac{3}{5} \times 10$ $10 \times \frac{3}{5}$

EUREKA MATH | Lesson 8:   Relate a fraction of a set to the repeated addition interpretation of fraction multiplication.

33

4. Solve each problem in two different ways as modeled in the example.

   Example: $6 \times \frac{2}{3} = \frac{6 \times 2}{3} = \frac{3 \times 2 \times 2}{3} = \frac{3 \times 4}{3} = 4$ $6 \times \frac{2}{3} = \frac{\overset{2}{\cancel{6}} \times 2}{\underset{1}{\cancel{3}}} = 4$

   a. $14 \times \frac{3}{7} =$ $14 \times \frac{3}{7}$

   b. $\frac{3}{4} \times 36$ $\frac{3}{4} \times 36$

   c. $30 \times \frac{13}{10}$ $30 \times \frac{13}{10}$

   d. $\frac{9}{8} \times 32$ $\frac{9}{8} \times 32$

5. Solve each problem any way you choose.

   a. $\frac{1}{2} \times 60$ $\frac{1}{2}$ minute = __30__ seconds

   b. $\frac{3}{4} \times 60$ $\frac{3}{4}$ hour = _____ minutes

   c. $\frac{3}{10} \times 1000$ $\frac{3}{10}$ kilogram = _____ grams

   d. $\frac{4}{5} \times 100$ $\frac{4}{5}$ meter = _____ centimeters

EUREKA
MATH™

Lesson 8:   Relate a fraction of a set to the repeated addition interpretation of
fraction multiplication.

34

Name _____ Date _____

1. Rewrite the following expressions as shown in the example.

Example: $\frac{2}{3} + \frac{2}{3} + \frac{2}{3} + \frac{2}{3} = \frac{4 \times 2}{3} = \frac{8}{3}$

a. $\frac{5}{3} + \frac{5}{3} + \frac{5}{3}$

b. $\frac{13}{5} + \frac{13}{5}$

c. $\frac{9}{4} + \frac{9}{4} + \frac{9}{4}$

2. Solve each problem in two different ways as modeled in the example.

Example: $\frac{2}{3} \times 6 = \frac{2 \times 6}{3} = \frac{12}{3} = 4$ $\frac{2}{3} \times 6 = \frac{2 \times \overset{2}{\cancel{6}}}{\cancel{3}_1} = 4$

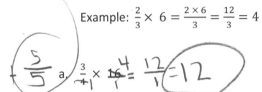

a. $\frac{3}{4} \times 16 = \frac{12}{1} = 12$ $\frac{3}{4} \times 16$

b. $\frac{4}{3} \times 12 = \frac{16}{1} = 16$ $\frac{4}{3} \times 12$

c. $40 \times \frac{11}{10} = \frac{44}{1} = 44$ $40 \times \frac{11}{10}$

d. $\frac{7}{6} \times 36 = \frac{42}{1} = 42$ $\frac{7}{6} \times 36$

e. $24 \times \frac{5}{8} = \frac{15}{1} = 15$ $24 \times \frac{5}{8}$

EUREKA MATH™

Lesson 8: Relate a fraction of a set to the repeated addition interpretation of fraction multiplication.

35

69

f. $\frac{18}{1} \times \frac{5}{12} = \frac{5 \times 3}{2 \times 1} = \frac{15}{2} = 7\frac{1}{2}$ $18 \times \frac{5}{12}$

g. $\frac{10}{9} \times \frac{21}{1} = \frac{16 \times 27}{3 \times 1} = \frac{76}{3} = 25\frac{1}{3}$ $\frac{10}{9} \times 21$

3. Solve each problem any way you choose.

a. $\frac{1}{3} \times 60$ $\frac{1}{3}$ minute = _____ seconds

b. $\frac{4}{5} \times 60$ $\frac{4}{5}$ hour = _____ minutes

c. $\frac{7}{10} \times 1000$ $\frac{7}{10}$ kilogram = _____ grams

d. $\frac{3}{5} \times 100$ $\frac{3}{5}$ meter = _____ centimeters

EUREKA
MATH™

Lesson 8:   Relate a fraction of a set to the repeated addition interpretation of
fraction multiplication.

36

Name _____   Date _____

1. Convert. Show your work using a tape diagram or an equation. The first one is done for you.

a. $\frac{1}{2}$ yard = ___$1\frac{1}{2}$___ feet

$\frac{1}{2}$ yard = $\frac{1}{2}$ × 1 yard

$= \frac{1}{2}$ × 3 feet

$= \frac{3}{2}$ feet

$= 1\frac{1}{2}$ feet

b. $\frac{1}{3}$ foot = _____ inches

$\frac{1}{3}$ foot = $\frac{1}{3}$ × 1 foot

$= \frac{1}{3}$ × 12 inches

$=$

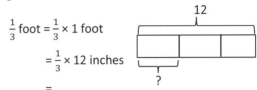

c. $\frac{5}{6}$ year = _____ months

d. $\frac{4}{5}$ meter = _____ centimeters

e. $\frac{2}{3}$ hour = _____ minutes

f. $\frac{3}{4}$ yard = _____ inches

Lesson 9:   Find a fraction of measurement, and solve word problems.

2. Mrs. Lang told her class that the class's pet hamster is $\frac{1}{4}$ ft in length. How long is the hamster in inches?

3. At the market, Mr. Paul bought $\frac{7}{8}$ lb of cashews and $\frac{3}{4}$ lb of walnuts.
    a. How many ounces of cashews did Mr. Paul buy?

    b. How many ounces of walnuts did Mr. Paul buy?

    c. How many more ounces of cashews than walnuts did Mr. Paul buy?

    d. If Mrs. Toombs bought $1\frac{1}{2}$ pounds of pistachios, who bought more nuts, Mr. Paul or Mrs. Toombs?
        How many ounces more?

4. A jewelry maker purchased 20 inches of gold chain. She used $\frac{3}{8}$ of the chain for a bracelet. How many
    inches of gold chain did she have left?

Lesson 9:   Find a fraction of measurement, and solve word problems.

38

Name _____   Date _____

1. Convert. Show your work using a tape diagram or an equation. The first one is done for you.

a. $\frac{1}{4}$ yard = _____9_____ inches
 	
$\frac{1}{4}$ yard = $\frac{1}{4}$ × 1 yard
 	
    = $\frac{1}{4}$ × 36 inches
 	
    = $\frac{36}{4}$ inches
 	
    = 9 inches
	b. $\frac{1}{6}$ foot = _____ inches
	
	$\frac{1}{6}$ foot = $\frac{1}{6}$ × 1 foot
	
	    = $\frac{1}{6}$ × 12 inches
	
	    =

c. $\frac{3}{4}$ year = _____ months	d. $\frac{3}{5}$ meter = _____ centimeters
e. $\frac{5}{12}$ hour = _____ minutes	f. $\frac{2}{3}$ yard = _____ inches

2. Michelle measured the length of her forearm. It was $\frac{3}{4}$ of a foot. How long is her forearm in inches?

3. At the market, Ms. Winn bought $\frac{3}{4}$ lb of grapes and $\frac{5}{8}$ lb of cherries.

   a. How many ounces of grapes did Ms. Winn buy?

   b. How many ounces of cherries did Ms. Winn buy?

   c. How many more ounces of grapes than cherries did Ms. Winn buy?

   d. If Mr. Phillips bought $1\frac{3}{4}$ pounds of raspberries, who bought more fruit, Ms. Winn or Mr. Phillips? How many ounces more?

4. A gardener has 10 pounds of soil. He used $\frac{5}{8}$ of the soil for his garden. How many pounds of soil did he use in the garden? How many pounds did he have left?

EUREKA
MATH™   Lesson 9:   Find a fraction of measurement, and solve word problems.

40

Name _____   Date _____

1. Write expressions to match the diagrams. Then, evaluate.

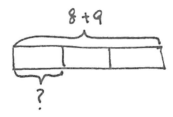

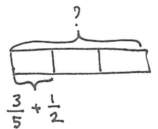

2. Write an expression to match, then evaluate.

   a. $\frac{1}{6}$ the sum of 16 and 20.

   b. Subtract 5 from $\frac{1}{3}$ of 23.

   c. 3 times as much as the sum of $\frac{3}{4}$ and $\frac{2}{6}$.

   d. $\frac{2}{5}$ of the product of $\frac{5}{6}$ and 42.

   e. 8 copies of the sum of 4 thirds and 2 more.

   f. 4 times as much as 1 third of 8.

3. Circle the expression(s) that gives the same product as $\frac{4}{5} \times 7$. Explain how you know.

$4 \div (7 \times 5)$ $7 \div 5 \times 4$ $(4 \times 7) \div 5$ $4 \div (5 \times 7)$ $4 \times \frac{7}{5}$ $7 \times \frac{4}{5}$

4. Use <, >, or = to make true number sentences without calculating. Explain your thinking.

a. $4 \times 2 + 4 \times \frac{2}{3}$ $\bigcirc$ $3 \times \frac{2}{3}$

b. $\left(5 \times \frac{3}{4}\right) \times \frac{2}{5}$ $\bigcirc$ $\left(5 \times \frac{3}{4}\right) \times \frac{2}{7}$

c. $3 \times \left(3 + \frac{15}{12}\right)$ $\bigcirc$ $(3 \times 3) + \frac{15}{12}$

EUREKA
MATH™   Lesson 10:   Compare and evaluate expressions with parentheses.

42

5. Collette bought milk for herself each month and recorded the amount in the table below. For (a–c), write an expression that records the calculation described. Then, solve to find the missing data in the table.

a. She bought $\frac{1}{4}$ of July's total in June.

Month	Amount (in gallons)
January	3
February	2
March	$1\frac{1}{4}$
April	
May	$\frac{7}{4}$
June	
July	2
August	1
September	
October	$\frac{1}{4}$

b. She bought $\frac{3}{4}$ as much in September as she did in January and July combined.

c. In April, she bought $\frac{1}{2}$ gallon less than twice as much as she bought in August.

d. Display the data from the table in a line plot.

e. How many gallons of milk did Collette buy from January to October?

EUREKA
MATH™   Lesson 10:   Compare and evaluate expressions with parentheses.

43

Name _____   Date _____

1. Write expressions to match the diagrams. Then, evaluate.

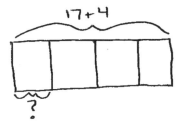

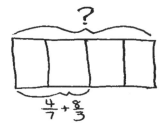

2. Circle the expression(s) that give the same product as $6 \times \frac{3}{8}$. Explain how you know.

$8 \div (3 \times 6)$ $3 \div 8 \times 6$ $(6 \times 3) \div 8$ $(8 \div 6) \times 3$ $6 \times \frac{8}{3}$ $\frac{3}{8} \times 6$

3. Write an expression to match, and then evaluate.

a. $\frac{1}{8}$ the sum of 23 and 17.

b. Subtract 4 from $\frac{1}{6}$ of 42.

c. 7 times as much as the sum of $\frac{1}{3}$ and $\frac{4}{5}$.

d. $\frac{2}{3}$ of the product of $\frac{3}{8}$ and 16.

e. 7 copies of the sum of 8 fifths and 4.

f. 15 times as much as 1 fifth of 12.

EUREKA MATH™   Lesson 10:   Compare and evaluate expressions with parentheses.

44

4. Use <, >, or = to make true number sentences without calculating. Explain your thinking.

   a. $\frac{2}{3} \times (9 + 12)$ ◯ $15 \times \frac{2}{3}$

   b. $\left(3 \times \frac{5}{4}\right) \times \frac{3}{5}$ ◯ $\left(3 \times \frac{5}{4}\right) \times \frac{3}{8}$

   c. $6 \times (2 + \frac{32}{16})$ ◯ $(6 \times 2) + \frac{32}{16}$

5. Fantine bought flour for her bakery each month and recorded the amount in the table to the right. For (a–c), write an expression that records the calculation described. Then, solve to find the missing data in the table.

   a. She bought $\frac{3}{4}$ of January's total in August.

   b. She bought $\frac{7}{8}$ as much in April as she did in October and July combined.

Month	Amount (in pounds)
January	3
February	2
March	$1\frac{1}{4}$
April	
May	$\frac{9}{8}$
June	
July	$1\frac{1}{4}$
August	
September	$\frac{11}{4}$
October	$\frac{3}{4}$

c. In June, she bought $\frac{1}{8}$ pound less than three times as much as she bought in May.

d. Display the data from the table in a line plot.

e. How many pounds of flour did Fantine buy from January to October?

Name _____   Date _____

1. Kim and Courtney share a 16-ounce box of cereal. By the end of the week, Kim has eaten $\frac{3}{8}$ of the box, and Courtney has eaten $\frac{1}{4}$ of the box of cereal. What fraction of the box is left?

2. Mathilde has 20 pints of green paint. She uses $\frac{2}{5}$ of it to paint a landscape and $\frac{3}{10}$ of it while painting a clover. She decides that, for her next painting, she will need 14 pints of green paint. How much more paint will she need to buy?

3. Jack, Jill, and Bill each carried a 48-ounce bucket full of water down the hill. By the time they reached the bottom, Jack's bucket was only $\frac{3}{4}$ full, Jill's was $\frac{2}{3}$ full, and Bill's was $\frac{1}{6}$ full. How much water did they spill altogether on their way down the hill?

4. Mrs. Diaz makes 5 dozen cookies for her class. One-ninth of her 27 students are absent the day she brings the cookies. If she shares the cookies equally among the students who are present, how many cookies will each student get?

5. Create a story problem about a fish tank for the tape diagram below. Your story must include a fraction.

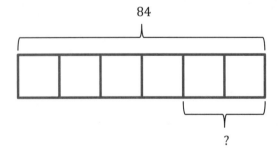

EUREKA
MATH™

Lesson 11:   Solve and create fraction word problems involving addition,
              subtraction, and multiplication..

48

Name _____   Date _____

1. Jenny's mom says she has an hour before it's bedtime. Jenny spends $\frac{1}{3}$ of the hour texting a friend and $\frac{1}{4}$ of the time brushing her teeth and putting on her pajamas. She spends the rest of the time reading her book. How many minutes did Jenny read?

2. A-Plus Auto Body is painting designs on a customer's car. They had 18 pints of blue paint on hand. They used $\frac{1}{2}$ of it for the flames, and $\frac{1}{3}$ of it for the sparks. They need $7\frac{3}{4}$ pints of blue paint to paint the next design. How many more pints of blue paint will they need to buy?

3. Giovanna, Frances, and their dad each carried a 10-pound bag of soil into the backyard. After putting soil in the first flower bed, Giovanna's bag was $\frac{5}{8}$ full, Frances' bag was $\frac{2}{5}$ full, and their dad's was $\frac{3}{4}$ full. How many pounds of soil did they put in the first flower bed altogether?

Lesson 11:   Solve and create fraction word problems involving addition, subtraction, and multiplication..

49

4. Mr. Chan made 252 cookies for the Annual Fifth Grade Class Bake Sale. They sold $\frac{3}{4}$ of them, and $\frac{3}{9}$ of the remaining cookies were given to P.T.A. members. Mr. Chan allowed the 12 student helpers to divide the cookies that were left equally. How many cookies will each student get?

5. Using the tape diagram below, create a story problem about a farm. Your story must include a fraction.

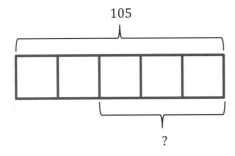

EUREKA
MATH

Lesson 11:   Solve and create fraction word problems involving addition, subtraction, and multiplication..

50

Name _____ Date _____

1. A baseball team played 32 games and lost 8. Katy was the catcher in $\frac{5}{8}$ of the winning games and $\frac{1}{4}$ of the losing games.
   a. What fraction of the games did the team win?

   b. In how many games did Katy play catcher?

2. In Mrs. Elliott's garden, $\frac{1}{8}$ of the flowers are red, $\frac{1}{4}$ of them are purple, and $\frac{1}{5}$ of the remaining flowers are pink. If there are 128 flowers, how many flowers are pink?

Lesson 12: Solve and create fraction word problems involving addition, subtraction, and multiplication.

51

3. Lillian and Darlene plan to get their homework finished within one hour. Darlene completes her math
    homework in $\frac{3}{5}$ hour. Lillian completes her math homework with $\frac{5}{6}$ hour remaining. Who completes her
    homework faster and by how many minutes?

    Bonus: Give the answer as a fraction of an hour.

4. Create and solve a story problem about a baker and some flour whose solution is given by the expression
    $\frac{1}{4} \times (3 + 5)$.

5. Create and solve a story problem about a baker and 36 kilograms of an ingredient that is modeled by the following tape diagram. Include at least one fraction in your story.

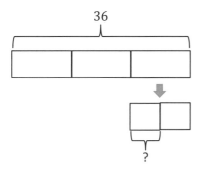

6. Of the students in Mr. Smith's fifth grade class, $\frac{1}{3}$ were absent on Monday. Of the students in Mrs. Jacobs' class, $\frac{2}{5}$ were absent on Monday. If there were 4 students absent in each class on Monday, how many students are in each class?

EUREKA
MATH™   Lesson 12:   Solve and create fraction word problems involving addition,
                             subtraction, and multiplication.   53

Name _____   Date _____

1. Terrence finished a word search in $\frac{3}{4}$ the time it took Frank. Charlotte finished the word search in $\frac{2}{3}$ the time it took Terrence. Frank finished the word search in 32 minutes. How long did it take Charlotte to finish the word search?

2. Ms. Phillips ordered 56 pizzas for a school fundraiser. Of the pizzas ordered, $\frac{2}{7}$ of them were pepperoni, 19 were cheese, and the rest were veggie pizzas. What fraction of the pizzas was veggie?

Lesson 12:   Solve and create fraction word problems involving addition,
               subtraction, and multiplication.

54

3. In an auditorium, $\frac{1}{6}$ of the students are fifth graders, $\frac{1}{3}$ are fourth graders, and $\frac{1}{4}$ of the remaining students are second graders. If there are 96 students in the auditorium, how many second graders are there?

4. At a track meet, Jacob and Daniel compete in the 220-m hurdles. Daniel finishes in $\frac{3}{4}$ of a minute. Jacob finishes with $\frac{5}{12}$ of a minute remaining. Who ran the race in the faster time?

Bonus: Express the difference in their times as a fraction of a minute.

5. Create and solve a story problem about a runner who is training for a race. Include at least one fraction in your story.

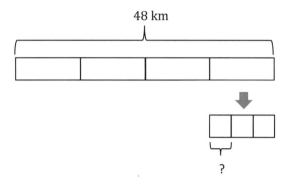

48 km

?

6. Create and solve a story problem about two friends and their weekly allowance whose solution is given by the expression $\frac{1}{5} \times (12 + 8)$.

EUREKA MATH | **Lesson 12:**   Solve and create fraction word problems involving addition, subtraction, and multiplication.

56

Name _____   Date _____

1. Solve. Draw a rectangular fraction model to show your thinking. Then, write a multiplication sentence. The first one has been done for you.

a. Half of $\frac{1}{4}$ pan of brownies = ___$\frac{1}{8}$___ pan of brownies

$$\frac{1}{2} \times \frac{1}{4} = \frac{1}{8}$$

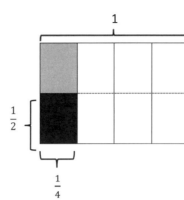

b. Half of $\frac{1}{3}$ pan of brownies = _____ pan of brownies

c. A fourth of $\frac{1}{3}$ pan of brownies = $\frac{1}{12}$ pan of brownies

$$\frac{1}{4} \times \frac{1}{3} = \frac{1}{12}$$

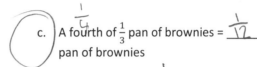
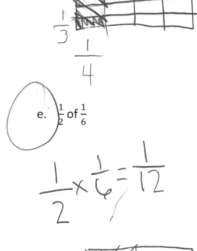

d. $\frac{1}{4}$ of $\frac{1}{4}$

$$\frac{1}{4} \times \frac{1}{4} = \frac{1}{16}$$

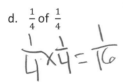
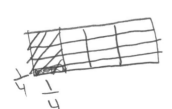

e. $\frac{1}{2}$ of $\frac{1}{6}$

$$\frac{1}{2} \times \frac{1}{6} = \frac{1}{12}$$

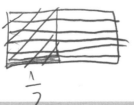

$$3 \times \tfrac{1}{4} > \tfrac{1}{3} \times \tfrac{1}{4}$$

2. Draw rectangular fraction models of $3 \times \frac{1}{4}$ and $\frac{1}{3} \times \frac{1}{4}$. Compare multiplying a number by 3 and by 1 third.

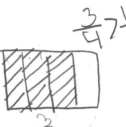
$$3 \times \tfrac{1}{4} = \tfrac{3}{4}$$

$$\tfrac{3}{4} > \tfrac{1}{2} \qquad \tfrac{1}{3} \times \tfrac{1}{4} = \tfrac{1}{12}$$

3. $\frac{1}{2}$ of Ila's workspace is covered in paper. $\frac{1}{3}$ of the paper is covered in yellow sticky notes. What fraction of Ila's workspace is covered in yellow sticky notes? Draw a picture to support your answer.

$$\tfrac{1}{2} \times \tfrac{1}{3} = \tfrac{1}{6} \text{ yellow sticky notes}$$

$\frac{1}{6}$ are yellow sticky notes.

4. A marching band is rehearsing in rectangular formation. $\frac{1}{5}$ of the marching band members play percussion instruments. $\frac{1}{2}$ of the percussionists play the snare drum. What fraction of all the band members play the snare drum?

$$\tfrac{1}{5} \times \tfrac{1}{2} = \tfrac{1}{10}$$

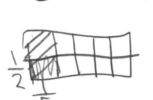

$\frac{1}{10}$ members play the snaredrum

5. Marie is designing a bedspread for her grandson's new bedroom. $\frac{2}{3}$ of the bedspread is covered in race cars and the rest is striped. $\frac{1}{4}$ of the stripes are red. What fraction of the bedspread is covered in red stripes?

EUREKA MATH™ | Lesson 13:   Multiply unit fractions by unit fractions.

58

Name _____ Date _____

1. Solve. Draw a rectangular fraction model to show your thinking.

a. Half of $\frac{1}{2}$ cake = __4__ cake

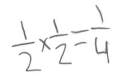

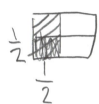

b. One-third of $\frac{1}{2}$ cake = __6__ cake

c. $\frac{1}{4}$ of $\frac{1}{2}$

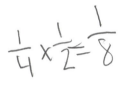

d. $\frac{1}{2} \times \frac{1}{5}$

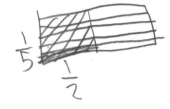

e. $\frac{1}{3} \times \frac{1}{3}$

f. $\frac{1}{4} \times \frac{1}{3}$

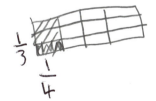

2. Noah mows $\frac{1}{2}$ of his property and leaves the rest wild. He decides to use $\frac{1}{5}$ of the wild area for a vegetable garden. What fraction of the property is used for the garden? Draw a picture to support your answer.

$$\frac{1}{2} \times \frac{1}{5} = \frac{1}{10}$$

$\frac{1}{10}$ of the property was used for the garden.

3. Fawn plants $\frac{2}{3}$ of the garden with vegetables. Her son plants the remainder of the garden. He decides to use $\frac{1}{2}$ of his space to plant flowers, and in the rest, he plants herbs. What fraction of the entire garden is planted in flowers? Draw a picture to support your answer.

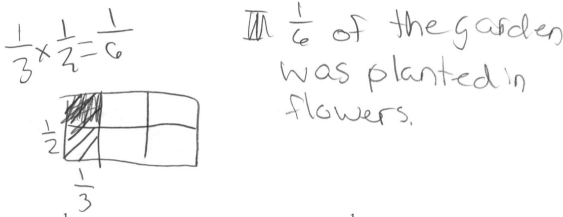

$$\frac{1}{3} \times \frac{1}{2} = \frac{1}{6}$$

$\frac{1}{6}$ of the garden was planted in flowers.

4. Diego eats $\frac{1}{5}$ of a loaf of bread each day. On Tuesday, Diego eats $\frac{1}{4}$ of the day's portion before lunch. What fraction of the whole loaf does Diego eat before lunch on Tuesday? Draw a rectangular fraction model to support your thinking.

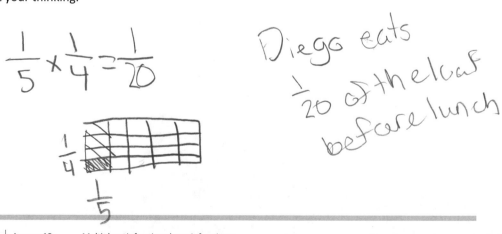

$$\frac{1}{5} \times \frac{1}{4} = \frac{1}{20}$$

Diego eats $\frac{1}{20}$ of the loaf before lunch

EUREKA MATH™   Lesson 13:   Multiply unit fractions by unit fractions.

60

Name _____   Date _____

1. Solve. Draw a rectangular fraction model to explain your thinking. Then, write a number sentence. An example has been done for you.

Example:

$\frac{1}{2}$ of $\frac{2}{5} = \frac{1}{2}$ of 2 fifths = 1 fifth

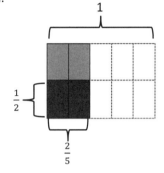

$\frac{1}{2} \times \frac{2}{5} = \frac{2}{10} = \frac{1}{5}$

a. $\frac{1}{3}$ of $\frac{3}{4} = \frac{1}{3}$ of _____ fourths = _____ fourth

b. $\frac{1}{2}$ of $\frac{4}{5} = \frac{1}{2}$ of _____ fifths = _____ fifths

c. $\frac{1}{2}$ of $\frac{2}{2} =$

d. $\frac{2}{3}$ of $\frac{1}{2} =$

e. $\frac{1}{2} \times \frac{3}{5} =$

f. $\frac{2}{3} \times \frac{1}{4} =$

2. $\frac{5}{8}$ of the songs on Harrison's music player are hip-hop. $\frac{1}{3}$ of the remaining songs are rhythm and blues.

    What fraction of all the songs are rhythm and blues? Use a tape diagram to solve.

3. Three-fifths of the students in a room are girls. One-third of the girls have blond hair. One-half of the boys have brown hair.
    a.   What fraction of all the students are girls with blond hair?

    b.   What fraction of all the students are boys without brown hair?

4. Cody and Sam mowed the yard on Saturday. Dad told Cody to mow $\frac{1}{4}$ of the yard. He told Sam to mow $\frac{1}{3}$ of the remainder of the yard. Dad paid each of the boys an equal amount. Sam said, "Dad, that's not fair! I had to mow one-third and Cody only mowed one-fourth!" Explain to Sam the error in his thinking. Draw a picture to support your reasoning.

Name _____   Date _____

1. Solve. Draw a rectangular fraction model to explain your thinking.

   a. $\frac{1}{2}$ of $\frac{2}{3}$ = $\frac{1}{2}$ of ____ thirds = ____ thirds

   b. $\frac{1}{2}$ of $\frac{4}{3}$ = $\frac{1}{2}$ of ____ thirds = ____ thirds

   c. $\frac{1}{3}$ of $\frac{3}{5}$ =

   d. $\frac{1}{2}$ of $\frac{6}{8}$ =

   e. $\frac{1}{3} \times \frac{4}{5}$ =

   f. $\frac{4}{5} \times \frac{1}{3}$ =

2. Sarah has a photography blog. $\frac{3}{7}$ of her photos are of nature. $\frac{1}{4}$ of the rest are of her friends. What fraction of all Sarah's photos is of her friends? Support your answer with a model.

EUREKA MATH™   Lesson 14:   Multiply unit fractions by non-unit fractions.

63

3. At Laurita's Bakery, $\frac{3}{5}$ of the baked goods are pies, and the rest are cakes. $\frac{1}{3}$ of the pies are coconut. $\frac{1}{6}$ of the cakes are angel-food.

    a. What fraction of all of the baked goods at Laurita's Bakery are coconut pies?

    b. What fraction of all of the baked goods at Laurita's Bakery are angel-food cakes?

4. Grandpa Mick opened a pint of ice cream. He gave his youngest grandchild $\frac{1}{5}$ of the ice cream and his middle grandchild $\frac{1}{4}$ of the remaining ice cream. Then, he gave his oldest grandchild $\frac{1}{3}$ of the ice cream that was left after serving the others.

    a. Who got the most ice cream? How do you know? Draw a picture to support your reasoning.

    b. What fraction of the pint of ice cream will be left if Grandpa Mick serves himself the same amount as the second grandchild?

EUREKA MATH   Lesson 14:   Multiply unit fractions by non-unit fractions.

64

Name _____　　Date _____

1. Solve. ~~Draw a rectangular fraction model to explain your thinking~~. Then, write a multiplication sentence. The first one is done for you.

a. $\frac{2}{3}$ of $\frac{3}{5}$

$$\frac{2}{3} \times \frac{3}{5} = \frac{6}{15} = \frac{2}{5}$$

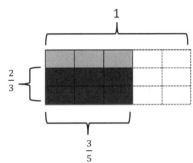

b. $\frac{3}{4}$ of $\frac{4}{5} = \frac{12 \cdot 4}{20 \cdot 4} = \frac{3}{5}$

$\frac{3}{4} \times \frac{4}{5} = \frac{3 \times 4}{4 \times 5} = \frac{3}{5}$

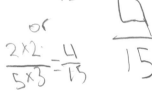

c. $\frac{2}{5}$ of $\frac{2}{3} = \frac{4}{15}$

or

$\frac{2 \times 2}{5 \times 3} = \frac{4}{15}$

d. $\frac{4}{5} \times \frac{2}{3} = \frac{8}{15}$

or

$\frac{4 \times 2}{5 \times 3} = \frac{8}{15}$

e. $\frac{3}{4} \times \frac{2}{3} = \frac{6 + 6}{12 \cdot 6} = \frac{1}{2}$

or

$\frac{3 \times 2}{4 \times 3} = \frac{1}{2}$

2. Multiply. ~~Draw a rectangular fraction model if it helps you~~, or use the method in the example.

Example: $\frac{6}{7} \times \frac{5}{8} = \frac{\overset{3}{\cancel{6}} \times 5}{7 \times \underset{4}{\cancel{8}}} = \frac{15}{28}$

a. $\frac{3}{4} \times \frac{5}{6}$

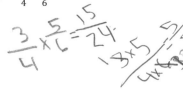

$\frac{3}{4} \times \frac{5}{6} = \frac{15}{24}$

$\frac{3 \times 5}{4 \times 2}$

b. $\frac{4}{5} \times \frac{5}{8}$

$\frac{4}{5} \times \frac{5}{8} = \frac{20}{40} = \frac{1}{2}$

$6 \times 12 = 72$ $14 \times 6 = 84$

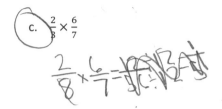

c. $\dfrac{2}{8} \times \dfrac{6}{7}$

$\dfrac{2}{8} \times \dfrac{6}{7} =$

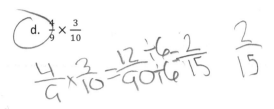

d. $\dfrac{4}{9} \times \dfrac{3}{10}$

$\dfrac{4}{9} \times \dfrac{3}{10} = \dfrac{12 \div 6}{90 \div 6} = \dfrac{2}{15}$ $\dfrac{2}{15}$

3. Phillip's family traveled $\dfrac{3}{10}$ of the distance to his grandmother's house on Saturday. They traveled $\dfrac{4}{7}$ of the remaining distance on Sunday. What fraction of the total distance to his grandmother's house was traveled on Sunday?

Phillip's
Family traveled
$\dfrac{2}{5}$ of the distance
on Sunday

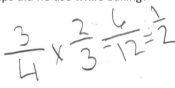

$\dfrac{7}{10}$

$\dfrac{7}{10}$ of $\dfrac{4}{7}$

$\dfrac{7}{10} \times \dfrac{4}{7}$ $\dfrac{4 \cdot 7 = 28 \div 7}{10 \div 7} = \dfrac{4 \cdot 2}{10 \cdot 2} = \dfrac{2}{5}$

4. Santino bought a $\dfrac{3}{4}$ pound bag of chocolate chips. He used $\dfrac{2}{3}$ of the bag while baking. How many pounds of chocolate chips did he use while baking?

$\dfrac{3}{4} \times \dfrac{2}{3} = \dfrac{6}{12} = \dfrac{1}{2}$ Santino bought $\dfrac{1}{2}$ pounds of the chocolate chips while baking.

5. Farmer Dave harvested his corn. He stored $\dfrac{5}{9}$ of his corn in one large silo and $\dfrac{3}{4}$ of the remaining corn in a small silo. The rest was taken to market to be sold.

a. What fraction of the corn was stored in the small silo?

$\dfrac{4}{9} \times \dfrac{3}{4} = \dfrac{12 \div 12}{36 \div 12} = \dfrac{1}{3}$ $\dfrac{1}{3}$ of the corn was the stored in the small silo

b. If he harvested 18 tons of corn, how many tons did he take to market?

$3 \times 12 = 36$
$3 \times 14 = 42$

Name _____   Date _____

1. Solve. Draw a rectangular fraction model to explain your thinking. Then, write a multiplication sentence.

a. $\frac{2}{3}$ of $\frac{3}{4} =$

$\frac{2}{3} \times \frac{3}{4} = \frac{6}{12} = \frac{1}{2}$ $\frac{1}{2}$

b. $\frac{2}{5}$ of $\frac{3}{4} = \frac{6 \div 2}{20 \div 2} = \frac{3}{10}$

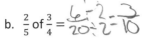

$\frac{3}{10}$

c. $\frac{2}{5}$ of $\frac{4}{5} = \frac{8}{25}$ $\frac{8}{25}$

d. $\frac{4}{5}$ of $\frac{3}{4} = \frac{12 \div 4}{20 \div 4} = \frac{3}{5}$ $\frac{3}{5}$

2. Multiply. Draw a rectangular fraction model if it helps you.

a. $\frac{5}{6} \times \frac{3}{10} = \frac{15}{60} ; \frac{15}{60 \div 15} = \frac{1}{4}$ $\frac{1}{4}$

b. $\frac{3}{4} \times \frac{4}{5} = \frac{12 \div 4}{20 \div 4} = \frac{3}{5}$ $\frac{3}{5}$

c. $\frac{5}{6} \times \frac{5}{8} = \frac{25}{48}$

$\frac{5 \times 5}{6 \times 8} = \frac{25}{48}$

d. $\frac{3}{4} \times \frac{5}{12} = \frac{15 \div 3}{48 \div 3} = \frac{5}{16}$

$\frac{3 \times 5}{4 \times 12 4} = \frac{5}{16}$ $\frac{5}{16}$

e. $\frac{8}{9} \times \frac{2}{3} = \frac{16}{27}$ $\frac{16}{27}$

f. $\frac{3}{7} \times \frac{2}{9} = \frac{2}{21}$ $\frac{2}{21}$

EUREKA MATH™ | **Lesson 15:**   Multiply non-unit fractions by non-unit fractions.

67

3. Every morning, Halle goes to school with a 1 liter bottle of water. She drinks $\frac{1}{4}$ of the bottle before school starts and $\frac{2}{3}$ of the rest before lunch.

   a. What fraction of the bottle does Halle drink after school starts, but before lunch?

   Halle drinks $\frac{1}{2}$ of the litter bottle.

   b. How many milliliters are left in the bottle at lunch?

   500 milliliters

4. Moussa delivered $\frac{3}{8}$ of the newspapers on his route in the first hour and $\frac{4}{5}$ of the rest in the second hour. What fraction of the newspapers did Moussa deliver in the second hour?

   Moussa delivered $\frac{1}{2}$ of the newspapers in the second hour.

5. Rose bought some spinach. She used $\frac{3}{5}$ of the spinach on a pan of spinach pie for a party, and $\frac{3}{4}$ of the remaining spinach for a pan for her family. She used the rest of the spinach to make a salad.

   a. What fraction of the spinach did she use to make the salad?

   b. If Rose used 3 pounds of spinach to make the pan of spinach pie for the party, how many pounds of spinach did Rose use to make the salad?

EUREKA MATH　|　Lesson 15:　　Multiply non-unit fractions by non-unit fractions.

68

Name _____   Date _____

Solve and show your thinking with a tape diagram.

1. Mrs. Onusko made 60 cookies for a bake sale. She sold $\frac{2}{3}$ of them and gave $\frac{3}{4}$ of the remaining cookies to the students working at the sale. How many cookies did she have left?

2. Joakim is icing 30 cupcakes. He spreads mint icing on $\frac{1}{5}$ of the cupcakes and chocolate on $\frac{1}{2}$ of the remaining cupcakes. The rest will get vanilla icing. How many cupcakes have vanilla icing?

3. The Booster Club sells 240 cheeseburgers. $\frac{1}{4}$ of the cheeseburgers had pickles, $\frac{1}{2}$ of the remaining burgers had onions, and the rest had tomato. How many cheeseburgers had tomato?

4. DeSean is sorting his rock collection. $\frac{2}{3}$ of the rocks are metamorphic and $\frac{3}{4}$ of the remainder are igneous rocks. If the 3 rocks left over are sedimentary, how many rocks does DeSean have?

5. Milan puts $\frac{1}{4}$ of her lawn-mowing money in savings and uses $\frac{1}{2}$ of the remaining money to pay back her sister. If she has $15 left, how much did she have at first?

6. Parks is wearing several rubber bracelets. $\frac{1}{3}$ of the bracelets are tie-dye, $\frac{1}{6}$ are blue, and $\frac{1}{3}$ of the remainder are camouflage. If Parks wears 2 camouflage bracelets, how many bracelets does he have on?

7. Ahmed spent $\frac{1}{3}$ of his money on a burrito and a water bottle. The burrito cost 2 times as much as the water. The burrito cost $4, how much money does Ahmed have left?

Name _____   Date _____

Solve and show your thinking with a tape diagram.

1. Anthony bought an 8-foot board. He cut off $\frac{3}{4}$ of the board to build a shelf, and gave $\frac{1}{3}$ of the rest to his brother for an art project. How many inches long was the piece Anthony gave to his brother?

2. Riverside Elementary School is holding a school-wide election to choose a school color. Five-eighths of the votes were for blue, $\frac{5}{9}$ of the remaining votes were for green, and the remaining 48 votes were for red.
   a. How many votes were for blue?

   b. How many votes were for green?

c. If every student got one vote, but there were 25 students absent on the day of the vote, how many students are there at Riverside Elementary School?

d. Seven-tenths of the votes for blue were made by girls. Did girls who voted for blue make up more than or less than half of all votes? Support your reasoning with a picture.

e. How many girls voted for blue?

Name _____ Date _____

1. Multiply and model. Rewrite each expression as a multiplication sentence with decimal factors. The first one is done for you.

a. $\frac{1}{10} \times \frac{1}{10}$

$= \frac{1 \times 1}{10 \times 10}$

$= \frac{1}{100}$

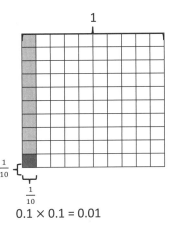

$0.1 \times 0.1 = 0.01$

b. $\frac{4}{10} \times \frac{3}{10}$

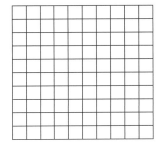

c. $\frac{1}{10} \times 1.4$

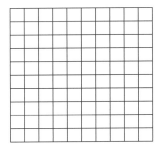 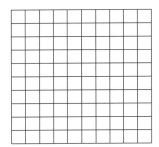

d. $\frac{6}{10} \times 1.7$

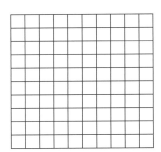 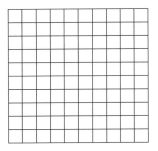

2. Multiply. The first few are started for you.

a. $5 \times 0.7 =$ 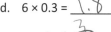 3.5

$= 5 \times \dfrac{7}{10}$

$= \dfrac{5 \times 7}{10}$

$= \dfrac{35}{10}$

$= 3.5$

b. $0.5 \times 0.7 =$ 0.35

$= \dfrac{5}{10} \times \dfrac{7}{10}$

$= \dfrac{5 \times 7}{10 \times 10}$

$= \dfrac{35}{100}$

c. $0.05 \times 0.7 =$ 0.035

$= \dfrac{5}{100} \times \dfrac{7}{10}$

$= \dfrac{5 \times 7}{100 \times 10}$

$= \dfrac{35}{1000}$

d. $6 \times 0.3 =$ 1.8

e. $0.6 \times 0.3 =$ 0.18

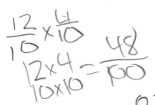

0.2

f. $0.06 \times 0.3 =$ 0.018

g. $1.2 \times 4 =$ 4.8

$\dfrac{12}{10} \times \dfrac{4}{1}$

$\dfrac{12 \times 4}{10 \times 1} = \dfrac{48}{10}$

h. $1.2 \times 0.4 =$ 0.48

$\dfrac{12}{10} \times \dfrac{4}{10}$

$\dfrac{12 \times 4}{10 \times 10} = \dfrac{48}{100}$

i. $0.12 \times 0.4 =$ 0.048

$\dfrac{12}{100} \times \dfrac{4}{10}$

$\dfrac{12 \times 4}{100 \times 10} \quad \dfrac{48}{1000}$

3. A boy scout has a length of rope measuring 0.7 meter. He uses 2 tenths of the rope to tie a knot at one end. How many meters of rope are in the knot?

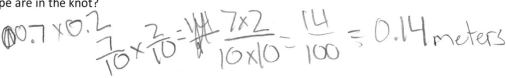

0.7×0.2 $\dfrac{7}{10} \times \dfrac{2}{10} = \dfrac{7 \times 2}{10 \times 10} = \dfrac{14}{100} = 0.14 \text{ meters}$

4. After just 4 tenths 0.4 of a 2.5 mile race was completed, Lenox took the lead and remained there until the end of the race.

a. How many miles did Lenox lead the race?

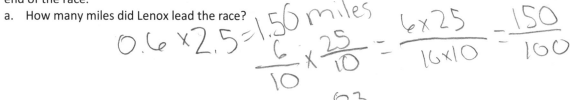

$0.6 \times 2.5 = 1.50 \text{ miles}$ $\dfrac{6}{10} \times \dfrac{25}{10} = \dfrac{6 \times 25}{16 \times 10} = \dfrac{150}{100}$

b. Reid, the second place finisher, developed a cramp with 3 tenths 0.3 of the race remaining. How many miles did Reid run without a cramp?

$0.7 \times 2.5 = 1.75 \text{ miles}$ $\dfrac{7}{10} \times \dfrac{25}{10} = \dfrac{7 \times 25}{16 \times 10} = \dfrac{175}{100} = 1.75$

Name _____ Date _____

1. Multiply and model. Rewrite each expression as a number sentence with decimal factors. The first one is done for you.

a. $\frac{1}{10} \times \frac{1}{10}$

$= \frac{1 \times 1}{10 \times 10}$

$= \frac{1}{100}$

0.1 × 0.1 = 0.01

b. $\frac{6}{10} \times \frac{2}{10}$

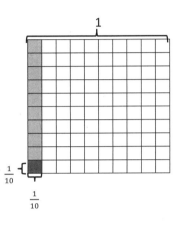

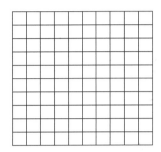

c. $\frac{1}{10} \times 1.6$

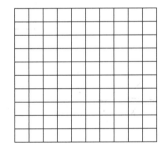
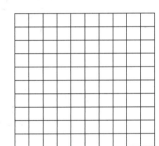

d. $\frac{6}{10} \times 1.9$

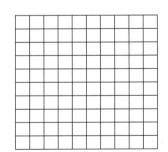
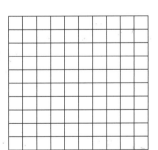

pg 74

2. Multiply. The first few are started for you.

a. $4 \times 0.6 = \underline{2.4}$

$= 4 \times \frac{6}{10}$

$= \frac{4 \times 6}{10}$

$= \frac{24}{10}$

$= 2.4$

b. $0.4 \times 0.6 = \underline{0.24}$

$= \frac{4}{10} \times \frac{6}{10}$

$= \frac{4 \times 6}{10 \times 10}$

$= \frac{24}{100}$

c. $0.04 \times 0.6 = \underline{0.024}$

$= \frac{4}{100} \times \frac{6}{10}$

$= \frac{4 \times 6}{100 \times 10}$

$= \frac{24}{1000}$

d. $7 \times 0.3 = \underline{2.1}$

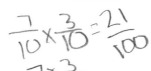

$7 \times \frac{3}{10}$

$\frac{7 \times 3}{1 \times 10} = \frac{21}{10}$

e. $0.7 \times 0.3 = \underline{0.21}$

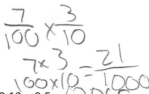

$\frac{7}{10} \times \frac{3}{10} = \frac{21}{100}$

$\frac{7 \times 3}{10 \times 10}$

f. $0.07 \times 0.3 = \underline{0.021}$

$\frac{7}{100} \times \frac{3}{10}$

$\frac{7 \times 3}{100 \times 10} = \frac{21}{1000}$

g. $1.3 \times 5 = \underline{6.5}$

$\begin{array}{r} 13 \\ \times 5 \\ \hline 6\,5 \end{array}$

$\frac{13}{10} \times \frac{5}{1}$

$\frac{13 \times 5}{10 \times 1} = \frac{65}{10}$

h. $1.3 \times 0.5 = \underline{0.65}$

$\frac{13}{10} \times \frac{5}{10} = \frac{65}{100} =$

$\frac{13 \times 5}{10 \times 10}$

i. $0.13 \times 0.5 = \underline{0.065}$

$\frac{13}{100} \times \frac{5}{10}$

$\frac{13 \times 5}{100 \times 10} = \frac{65}{1000}$

3. Jennifer makes 1.7 liters of lemonade. If she pours 3 tenths of the lemonade in the glass, how many liters of lemonade are in the glass?

$\frac{17}{10} \times \frac{3}{10} = \frac{17 \times 3}{10 \times 10} = \frac{61}{100} = 0.51$ liters

4. Cassius walked 6 tenths of a 3.6 mile trail. 0.6

a. How many miles did Cassius have left to hike?

0.4×3.6

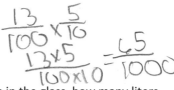

$\frac{4}{10} \times \frac{36}{10} = \frac{144}{100} = 1.44$

Cassius had 1.44 miles to hike

b. Cameron was 1.3 miles ahead of Cassius. How many miles did Cameron hike already?

0.6×3.6

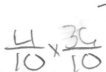

$\frac{6}{10} \times \frac{36}{10} = \frac{6 \times 36}{10 \times 10} = \frac{216}{100}$

Cameron hiked 3.46 miles.

$\begin{array}{r} 2.16 \\ + 1.30 \\ \hline 3.46 \end{array}$

1,000,000	100,000	10,000	1,000	100	10	1	.	$\frac{1}{10}$	$\frac{1}{100}$	$\frac{1}{1000}$
Millions	Hundred Thousands	Ten Thousands	Thousands	Hundreds	Tens	Ones	.	Tenths	Hundredths	Thousandths
							.			
							.			
							.			
							.			
							.			
							.			
							.			
							.			
							.			

millions through thousandths place value chart

Lesson 17:   Relate decimal and fraction multiplication.

77

Name _____   Date _____

1. Multiply using both fraction form and unit form. Check your answer by counting the decimal places. The first one is done for you.

   a. $2.3 \times 1.8 = \frac{23}{10} \times \frac{18}{10}$

   $$= \frac{23 \times 18}{100}$$

   $$= \frac{414}{100}$$

   $$= 4.14$$

$$
\begin{array}{r}
2\ 3 \text{ tenths} \\
\times \quad 1\ 8 \text{ tenths} \\
\hline
1\ 8\ 4 \\
+ \quad 2\ 3\ 0 \\
\hline
4\ 1\ 4 \text{ hundredths}
\end{array}
$$

   b. $2.3 \times 0.9 =$

$$
\begin{array}{r}
2\ 3 \text{ tenths} \\
\times \qquad 9 \text{ tenths} \\
\hline
\end{array}
$$

   c. $6.6 \times 2.8 =$

   d. $3.3 \times 1.4 =$

2. Multiply using fraction form and unit form. Check your answer by counting the decimal places. The first one is done for you.

   a. $2.38 \times 1.8 = \frac{238}{100} \times \frac{18}{10}$

   $$= \frac{238 \times 18}{1,000}$$

   $$= \frac{4,284}{1,000}$$

   $$= 4.284$$

$$
\begin{array}{r}
2\ 3\ 8 \text{ hundredths} \\
\times \qquad 1\ 8 \text{ tenths} \\
\hline
1\ 9\ 0\ 4 \\
+ \quad 2\ 3\ 8\ 0 \\
\hline
4,2\ 8\ 4 \text{ thousandths}
\end{array}
$$

   b. $2.37 \times 0.9 =$

$$
\begin{array}{r}
2\ 3\ 7 \text{ hundredths} \\
\times \qquad 9 \text{ tenths} \\
\hline
\end{array}
$$

   c. $6.06 \times 2.8 =$

   d. $3.3 \times 0.14 =$

2. Solve using the standard algorithm. Show your thinking about the units of your product. The first one is done for you.

   a. 3.2 × 0.6 = 1.92

       3 2 tenths $\dfrac{32}{10} \times \dfrac{6}{10} = \dfrac{32 \times 6}{100}$
   ×   6 tenths
     1 9 2 hundredths

   b. 3.2 × 1.2 = _____

       3 2 tenths
   ×   1 2 tenths

   c. 8.31 × 2.4 = _____

   d. 7.50 × 3.5 = _____

3. Carolyn buys 1.2 pounds of chicken breast. If each pound of chicken breast costs $3.70, how much will she pay for the chicken breast?

4. A kitchen measures 3.75 meters by 4.2 meters.
   a. Find the area of the kitchen.

   b. The area of the living room is one and a half times that of the kitchen. Find the total area of the living room and the kitchen.

Name _____   Date _____

1. Multiply using fraction form and unit form. Check your answer by counting the decimal places. The first one is done for you.

   a. $3.3 \times 1.6 = \frac{33}{10} \times \frac{16}{10}$

   $= \frac{33 \times 16}{100}$

   $= \frac{528}{100}$

   $= 5.28$

   ```
              3 3 tenths
        ×     1 6 tenths
              1 9 8
      +   3 3 0
            5 2 8 hundredths
   ```

   b. $3.3 \times 0.8 =$

   ```
              3 3 tenths
        ×        8 tenths
   ```

   c. $4.4 \times 3.2 =$

   d. $2.2 \times 1.6 =$

3. Multiply. The first one is partially done for you.

   a. $3.36 \times 1.4 = \frac{336}{100} \times \frac{14}{10}$

   $= \frac{336 \times 14}{1,000}$

   $= \frac{4,704}{1,000}$

   $= 4.704$

   ```
              3 3 6 hundredths
        ×         1 4 tenths
   ```

   b. $3.35 \times 0.7 =$

   ```
              3 3 5 hundredths
        ×            7 tenths
   ```

   c. $4.04 \times 3.2 =$

   d. $4.4 \times 0.16 =$

4. Solve using the standard algorithm. Show your thinking about the units of your product. The first one is done for you.

a. 3.2 × 0.6 = 1.92

$$\frac{32}{10} \times \frac{6}{10} = \frac{32 \times 6}{100}$$

     3 2 tenths
×   6 tenths
   1 9 2 hundredths

b. 2.3 × 2.1 = _____

     2 3 tenths
×   2 1 tenths

c. 7.41 × 3.4 = _____

d. 6.50 × 4.5 = _____

5. Erik buys 2.5 pounds of cashews. If each pound of cashews costs $7.70, how much will he pay for the cashews?

6. A swimming pool at a park measures 9.75 meters by 7.2 meters.
   a. Find the area of the swimming pool.

   b. The area of the playground is one and a half times that of the swimming pool. Find the total area of the swimming pool and the playground.

Name _____ Date _____

1. Convert. Express your answer as a mixed number, if possible. The first one is done for you.

a. 2 ft = ___$\frac{2}{3}$___ yd
 	
  2 ft = 2 × 1 ft
 	
  = 2 × $\frac{1}{3}$ yd
 	
  = $\frac{2}{3}$ yd
	b. 4 ft = _____ yd
	
	  4 ft = 4 × 1 ft
	
	  = 4 × _____ yd
	
	  = _____ yd
	
	  =

c. 7 in = _____ ft	d. 13 in = _____ ft
e. 5 oz = _____ lb	f. 18 oz = _____ lb

Lesson 19: Convert measures involving whole numbers, and solve multi-step worc problems.

83

2. Regina buys 24 inches of trim for a craft project.
    a. What fraction of a yard does Regina buy?

    b. If a whole yard of trim costs $6, how much did Regina pay?

3. At Yo-Yo Yogurt, the scale says that Sara has 8 ounces of vanilla yogurt in her cup. Her father's yogurt
    weighs 11 ounces. How many pounds of frozen yogurt did they buy altogether? Express your answer as a
    mixed number.

4. Pheng-Xu drinks 1 cup of milk every day for lunch. How many gallons of milk does he drink in 2 weeks?

Lesson 19:   Convert measures involving whole numbers, and solve multi-step worc
              problems.

84

Name _____ Date _____

1. Convert. Express your answer as a mixed number, if possible.

a. 2 ft = _____$\frac{2}{3}$_____ yd 2 ft = 2 × 1 ft = 2 × $\frac{1}{3}$ yd = $\frac{2}{3}$ yd	b. 6 ft = _____ yd 6 ft = 6 × 1 ft = 6 × _____ yd = _____ yd
c. 5 in = _____ ft	d. 14 in = _____ ft
e. 7 oz = _____ lb	f. 20 oz = _____ lb
g. 1 pt = _____ qt	h. 4 pt = _____ qt

Lesson 19: Convert measures involving whole numbers, and solve multi-step word problems.

85

2. Marty buys 12 ounces of granola.

   a. What fraction of a pound of granola did Marty buy?

   b. If a whole pound of granola costs $4, how much did Marty pay?

3. Sara and her dad visit Yo-Yo Yogurt again. This time, the scale says that Sara has 14 ounces of vanilla yogurt in her cup. Her father's yogurt weighs half as much. How many pounds of frozen yogurt did they buy altogether on this visit? Express your answer as a mixed number.

4. An art teacher uses 1 quart of blue paint each month. In one year, how many gallons of paint will she use?

Lesson 19: Convert measures involving whole numbers, and solve multi-step word problems.

86

Name _____ Date _____

1. Convert. Show your work. Express your answer as a mixed number. (Draw a tape diagram if it helps you.) The first one is done for you.

a. $2\frac{2}{3}$ yd = __8__ ft $2\frac{2}{3}$ yd $= 2\frac{2}{3} \times 1$ yd $\quad = 2\frac{2}{3} \times 3$ ft $\quad = \frac{8}{3} \times 3$ ft $\quad = \frac{24}{3}$ ft $\quad = 8$ ft	b. $1\frac{1}{2}$ qt = _____ gal $1\frac{1}{2}$ qt $= 1\frac{1}{2} \times 1$ qt $\quad = 1\frac{1}{2} \times \frac{1}{4}$ gal $\quad = \frac{3}{2} \times \frac{1}{4}$ gal $\quad =$
c. $4\frac{2}{3}$ ft = _____ in	d. $9\frac{1}{2}$ pt = _____ qt
e. $3\frac{3}{5}$ hr = _____ min	f. $3\frac{2}{3}$ ft = _____ yd

EUREKA MATH

Lesson 20: Convert mixed unit measurements, and solve multi-step word problems.

87

2. Three dump trucks are carrying topsoil to a construction site. Truck A carries 3,545 lb, Truck B carries 1,758 lb, and Truck C carries 3,697 lb. How many tons of topsoil are the 3 trucks carrying altogether?

3. Melissa buys $3\frac{3}{4}$ gallons of iced tea. Denita buys 7 quarts more than Melissa. How much tea do they buy altogether? Express your answer in quarts.

4. Marvin buys a hose that is $27\frac{3}{4}$ feet long. He already owns a hose at home that is $\frac{2}{3}$ the length of the new hose. How many total yards of hose does Marvin have now?

Lesson 20:   Convert mixed unit measurements, and solve multi-step word problems.

88

Name _____   Date _____

1. Convert. Show your work. Express your answer as a mixed number. The first one is done for you.

a. $2\frac{2}{3}$ yd = __8__ ft

$2\frac{2}{3}$ yd $= 2\frac{2}{3} \times 1$ yd

$\qquad = 2\frac{2}{3} \times 3$ ft

$\qquad = \frac{8}{3} \times 3$ ft

$\qquad = \frac{24}{3}$ ft

$\qquad = 8$ ft

b. $1\frac{1}{4}$ ft = _____ yd

$1\frac{1}{4}$ ft $= 1\frac{1}{4} \times 1$ ft

$\qquad = 1\frac{1}{4} \times \frac{1}{3}$ yd

$\qquad = \frac{5}{4} \times \frac{1}{3}$ yd

$\qquad =$

c. $3\frac{5}{6}$ ft = _____ in

d. $7\frac{1}{2}$ pt = _____ qt

e. $4\frac{3}{10}$ hr = _____ min

f. 33 months = _____ years

2. Four members of a track team run a relay race in 165 seconds. How many minutes did it take them to run the race?

3. Horace buys $2\frac{3}{4}$ pounds of blueberries for a pie. He needs 48 ounces of blueberries for the pie. How many more pounds of blueberries does he need to buy?

4. Tiffany is sending a package that may not exceed 16 pounds. The package contains books that weigh a total of $9\frac{3}{8}$ pounds. The other items to be sent weigh $\frac{3}{5}$ the weight of the books. Will Tiffany be able to send the package?

Lesson 20: Convert mixed unit measurements, and solve multi-step word problems.

90

Name _____   Date _____

1. Fill in the blanks. The first one has been done for you.

   a. $\frac{1}{4} \times 1 = \frac{1}{4} \times \frac{3}{3} = \frac{3}{12}$

   b. $\frac{3}{4} \times 1 = \frac{3}{4} \times \underline{} = \frac{21}{28}$

   c. $\frac{7}{4} \times 1 = \frac{7}{4} \times \underline{} = \frac{35}{20}$

   d. Use words to compare the size of the product to the size of the first factor.

2. Express each fraction as an equivalent decimal.

   a. $\frac{1}{4} \times \frac{25}{25} =$

   b. $\frac{3}{4} \times \frac{25}{25} =$

   c. $\frac{1}{5} \times \underline{} =$

   d. $\frac{4}{5} \times \underline{} =$

   e. $\frac{1}{20}$

   f. $\frac{27}{20}$

   g. $\frac{7}{4}$

   h. $\frac{8}{5}$

   i. $\frac{24}{25}$

   j. $\frac{93}{50}$

   k. $2\frac{6}{25}$

   l. $3\frac{31}{50}$

EUREKA
MATH™

Lesson 21: Explain the size of the product, and relate fraction and decimal
equivalence to multiplying a fraction by 1.

91

3. Jack said that if you take a number and multiply it by a fraction, the product will always be smaller than what you started with. Is he correct? Why or why not? Explain your answer, and give at least two examples to support your thinking.

4. There is an infinite number of ways to represent 1 on the number line. In the space below, write at least four expressions multiplying by 1. Represent *one* differently in each expression.

5. Maria multiplied by 1 to rename $\frac{1}{4}$ as hundredths. She made factor pairs equal to 10. Use her method to change one-eighth to an equivalent decimal.

$$\text{Maria's way:}\quad \frac{1}{4} = \frac{1}{2 \times 2} \times \frac{5 \times 5}{5 \times 5} = \frac{5 \times 5}{(2 \times 5) \times (2 \times 5)} = \frac{25}{100} = 0.25$$

$$\frac{1}{8} =$$

Paulo renamed $\frac{1}{8}$ as a decimal, too. He knows the decimal equal to $\frac{1}{4}$, and he knows that $\frac{1}{8}$ is half as much as $\frac{1}{4}$. Can you use his ideas to show another way to find the decimal equal to $\frac{1}{8}$?

EUREKA
MATH™

Lesson 21:   Explain the size of the product, and relate fraction and decimal
              equivalence to multiplying a fraction by 1.

92

Name _____   Date _____

1. Fill in the blanks.

   a. $\frac{1}{3} \times 1 = \frac{1}{3} \times \frac{3}{3} = \frac{}{9}$

   b. $\frac{2}{3} \times 1 = \frac{2}{3} \times \frac{}{} = \frac{14}{21}$

   c. $\frac{5}{2} \times 1 = \frac{5}{2} \times \frac{}{} = \frac{25}{}$

   d. Compare the first factor to the value of the product.

2. Express each fraction as an equivalent decimal. The first one is partially done for you.

   a. $\frac{3}{4} \times \frac{25}{25} = \frac{3 \times 25}{4 \times 25} = \frac{}{100} = $

   b. $\frac{1}{4} \times \frac{25}{25} = $

   c. $\frac{2}{5} \times \frac{}{} = $

   d. $\frac{3}{5} \times \frac{}{} = $

   e. $\frac{3}{20}$

   f. $\frac{25}{20}$

EUREKA MATH

Lesson 21:   Explain the size of the product, and relate fraction and decimal
equivalence to multiplying a fraction by 1.

93

g. $\frac{23}{25}$

h. $\frac{89}{50}$

i. $3\frac{11}{25}$

j. $5\frac{41}{50}$

3. $\frac{6}{8}$ is equivalent to $\frac{3}{4}$. How can you use this to help you write $\frac{6}{8}$ as a decimal? Show your thinking to solve.

4. A number multiplied by a fraction is not always smaller than the original number. Explain this and give at least two examples to support your thinking.

5. Elise has $\frac{3}{4}$ of a dollar. She buys a stamp that costs 44 cents. Change both numbers into decimals, and tell how much money Elise has after paying for the stamp.

EUREKA MATH   Lesson 21: Explain the size of the product, and relate fraction and decimal equivalence to multiplying a fraction by 1.

94

Name _____   Date _____

1. Solve for the unknown. Rewrite each phrase as a multiplication sentence. Circle the scaling factor and put a box around the number of meters.

   a. $\frac{1}{2}$ as long as 8 meters = ____4____ meters

   b. 8 times as long as $\frac{1}{2}$ meter = ____4____ meters

2. Draw a tape diagram to model each situation in Problem 1, and describe what happened to the number of meters when it was multiplied by the scaling factor.

   a.   b.

3. Fill in the blank with a numerator or denominator to make the number sentence true.

   a. $7 \times \frac{3}{4} < 7$ b. $\frac{7}{6} \times 15 > 15$ c. $3 \times \frac{5}{5} = 3$

4. Look at the inequalities in each box. Choose a single fraction to write in all three blanks that would make all three number sentences true. Explain how you know.

   a.

$\frac{3}{4} \times \frac{4}{3} > \frac{3}{4}$	$2 \times \frac{4}{3} > 2$	$\frac{7}{5} \times \frac{4}{3} > \frac{7}{5}$

   The scaling factor $\frac{4}{3}$ is greater than one.
   The product is greater than the factor

   b.

$\frac{3}{4} \times \frac{2}{3} < \frac{3}{4}$	$2 \times \frac{2}{3} < 2$	$\frac{7}{5} \times \frac{2}{3} < \frac{7}{5}$

   The scaling factor $\frac{2}{3}$ is less than 1.
   The product is less than the factor,

EUREKA MATH

Lesson 22:   Compare the size of the product to the size of the factors.

95

5. Johnny says multiplication always makes numbers bigger. Explain to Johnny why this isn't true.
   Give more than one example to help him understand.

6. A company uses a sketch to plan an advertisement on the side of a building. The lettering on the sketch is
   $\frac{3}{4}$ inch tall. In the actual advertisement, the letters must be 34 times as tall. How tall will the letters be on
   the building?

   $\frac{2}{3}$ in
   $\times \frac{3}{51}$

   34 × $\frac{3}{4}$ $\frac{51 \text{in}}{2} = 25\frac{1}{2}$ inches tall

   The letters will be $25\frac{1}{2}$ tall

7. Jason is drawing the floor plan of his bedroom. He is drawing everything with dimensions that are $\frac{1}{12}$ of
   the actual size. His bed measures 6 ft by 3 ft, and the room measures 14 ft by 16 ft. What are the
   dimensions of his bed and room in his drawing?

EUREKA
MATH

Lesson 22:   Compare the size of the product to the size of the factors.

96

Name _____   Date _____

1. Solve for the unknown. Rewrite each phrase as a multiplication sentence. Circle the scaling factor and put a box around the number of meters.

    a. $\frac{1}{3}$ as long as 6 meters = _____ meters

    b. 6 times as long as $\frac{1}{3}$ meter = _____ meters

2. Draw a tape diagram to model each situation in Problem 1, and describe what happened to the number of meters when it was multiplied by the scaling factor.

    a.

    b.

3. Fill in the blank with a numerator or denominator to make the number sentence true.

    a. $5 \times \frac{5}{3} > 9$

    b. $\frac{6}{7} \times 12 < 13$

    c. $4 \times \frac{5}{5} = 4$

4. Look at the inequalities in each box. Choose a single fraction to write in all three blanks that would make all three number sentences true. Explain how you know.

    a.
$\frac{2}{3} \times \frac{4}{3} > \frac{2}{3}$	$4 \times \frac{4}{3} > 4$	$\frac{5}{3} \times \frac{4}{3} > \frac{5}{3}$

    The scaling factor $\frac{4}{3}$ is greater than 1
    The product is greater than the factor

    b.
$\frac{2}{3} \times \frac{1}{3} < \frac{2}{3}$	$4 \times \frac{1}{3} < 4$	$\frac{5}{3} \times \frac{1}{3} < \frac{5}{3}$

    The scaling factor $\frac{1}{3}$ is less than 1
    The product is less than the factor.

EUREKA MATH™   Lesson 22:   Compare the size of the product to the size of the factors.

97

5. Write a number in the blank that will make the number sentence true.

a. $3 \times$ _____ < 1

b. Explain how multiplying by a whole number can result in a product less than 1.

6. In a sketch, a fountain is drawn $\frac{1}{4}$ yard tall. The actual fountain will be 68 times as tall. How tall will the fountain be?

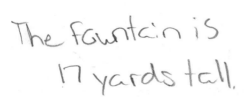

The fountain is
17 yards tall.

7. In blueprints, an architect's firm drew everything $\frac{1}{24}$ of the actual size. The windows will actually measure 4 ft by 6 ft and doors measure 12 ft by 8 ft. What are the dimensions of the windows and the doors in the drawing?

EUREKA
MATH™

Lesson 22:   Compare the size of the product to the size of the factors.

98

_265_

Name _____   Date _____

1. Fill in the blank using one of the following scaling factors to make each number sentence true.

1.021	0.989	1.00

   a. $3.4 \times \underline{1.00} = 3.4$ b. $\underline{1.021} \times 0.21 > 0.21$ c. $8.04 \times \underline{0.989} < 8.04$

2.

   a. Sort the following expressions by rewriting them in the table.

The product is less than the boxed number:	The product is greater than the boxed number:
0.3×0.069 602×0.489 0.2×0.1	13.89×1.004 102.03×4.015 0.72×1.24

factor   scaling

$\boxed{13.89} \times 1.004$ factor ✓ $\boxed{602} \times 0.489$ ✓ $\boxed{102.03} \times 4.015$ ✓

$\boxed{0.3} \times 0.069$ ✓ $\boxed{0.72} \times 1.24$ ✓ $\boxed{0.2} \times 0.1$ ✓

   b. Explain your sorting by writing a sentence that tells what the expressions in each column of the table have in common.

Less than	Greater than
The scaling factor is less than one. The product is less than the factor	The scaling factor is greater than one. The product is greater than the factor

3. Write a statement using one of the following phrases to compare the value of the expressions.
    Then, explain how you know.

    is slightly more than   is a lot more than   is slightly less than   is a lot less than

    a. 4×0.988 <u>is slightly less than</u> $\overset{F}{4}$ 0.988 is slightly less than one

    b. $\overset{SF}{1.05} \times \overset{F}{0.8}$ <u>slightly more than</u> $\overset{F}{0.8}$ 1.05 is slightly more than 1

    c. $\overset{F}{1,725} \times \overset{SF}{0.013}$ <u>a lot less than</u>   1,725 0.013 is a lot less than 1

    d. $\overset{SF}{989.001} \times \overset{F}{1.003}$ <u>lot more than</u>   1.003 989.001 is a lot more than 1

    e. $\overset{F}{0.002} \times \overset{SF}{0.911}$ <u>slightly less than</u>   0.002 0.911 is slightly less than 1

4. During science class, Teo, Carson, and Dhakir measure the length of their bean sprouts. Carson's sprout is
    0.9 times the length of Teo's, and Dhakir's is 1.08 times the length of Teo's. Whose bean sprout is the
    longest? The shortest? Explain your reasoning.

    Teo   [Bean sprout]
    Carson [0.9 times] ← Carson
    Dhakir [1.08 times] ← Longest

5. Complete the following statements, then use decimals to give an example of each.

    ▪ $a \times b > a$ will always be true when b is…

    $a \times b > a$

    $2.036 \times 1.086 > 2.036$

    ▪ $a \times b < a$ will always be true when b is…

    $A \times B < a$

    $999.9 \times 0.99 < 999.9$

Name _____   Date _____

1.
   a. Sort the following expressions by rewriting them in the table.

The product is less than the boxed number:	The product is greater than the boxed number:
828×0.921 0.05×0.1	12.5×1.989 0.007×1.62 2.16×1.11 321.46×1.26

$\boxed{12.5}$ × 1.989	$\boxed{828}$ × 0.921	$\boxed{321.46}$ × 1.26
$\boxed{0.007}$ × 1.02	$\boxed{2.16}$ × 1.11	$\boxed{0.05}$ × 0.1

b. What do the expressions in each column have in common?

Study

Less than	More than
The scaling Factor is less than 1. The product is	The scaling factor is more than 1. The product is more than 1. the product

one
the
product

2. Write a statement using one of the following phrases to compare the value of the expressions. Then, explain how you know.

   is slightly more than   *is a lot more than*   *is slightly less than*   *is a lot less than*

   a. SF
       14 × 0.999   ____slightly less than____ F 14   0.999 is slightly less than 1

   b. SF
       1.01 × 2.06 ____slightly more than____ 2.06   1.01 is slightly more than 1.

   c. SF
       1,955 × 0.019 ____lot less than____ 1,955   0.019 is a lot slightly less than 1

EUREKA MATH

Lesson 23:   Compare the size of the product to the size of the factors.

101

d.   Two thousand × 1.0001 $\underset{SF}{\underline{\text{slightly more than}}}$ two thousand   1.0001 is
slightly ~~is~~ more
than 1.

e.   Two-thousandths × 0.911 $\underset{SF}{\underline{\text{slightly less than}}}$ two-thousandths   0.911 is slightly
less than 1

3.   Rachel is 1.5 times as heavy as her cousin, Kayla. Another cousin, Jonathan, weighs 1.25 times as much as Kayla. List the cousins, from lightest to heaviest, and explain your thinking.

Kayla [   ] lightest
Jonathan [1.25 times]
Rachel [1.5 times] Heaviest

4.   Circle your choice.

a. $a \times b > a$

For this statement to be true, b must be   (greater than 1)   less than 1

Write two expressions that support your answer. Be sure to include one decimal example.

A × B > a
3.3 × 3.3 > 3.3

A × B > A
2.4 × 1.6 > 2.4

b. $a \times b < a$

For this statement to be true, b must be   greater than 1   (less than 1)

Write two expressions that support your answer. Be sure to include one decimal example.

A   B
6.5 × 0.3 < 6.5

2.3 × 0.9 < 2.3

EUREKA MATH   Lesson 23:   Compare the size of the product to the size of the factors.

102

Name _____ Date _____

1. A vial contains 20 mL of medicine. If each dose is $\frac{1}{8}$ of the vial, how many mL is each dose? Express your answer as a decimal.

2. A container holds 0.7 liters of oil and vinegar. $\frac{3}{4}$ of the mixture is vinegar. How many liters of vinegar are in the container? Express your answer as both a fraction and a decimal.

3. Andres completed a 5-km race in 13.5 minutes. His sister's time was $1\frac{1}{2}$ times longer than his time. How long, in minutes, did it take his sister to run the race?

Lesson 24: Solve word problems using fraction and decimal multiplication.

103

4. A clothing factory uses 1,275.2 meters of cloth a week to make shirts. How much cloth is needed to make $3\frac{3}{5}$ times as many shirts?

5. There are $\frac{3}{4}$ as many boys as girls in a class of fifth-graders. If there are 35 students in the class, how many are girls?

6. Ciro purchased a concert ticket for \$56. The cost of the ticket was $\frac{4}{5}$ the cost of his dinner. The cost of his hotel was $2\frac{1}{2}$ times as much as his ticket. How much did Ciro spend altogether for the concert ticket, hotel, and dinner?

Name _____ Date _____

1. Jesse takes his dog and cat for their annual vet visit. Jesse's dog weighs 23 pounds. The vet tells him his cat's weight is $\frac{5}{8}$ as much as his dog's weight. How much does his cat weigh?

2. An image of a snowflake is 1.8 centimeters wide. If the actual snowflake is $\frac{1}{8}$ the size of the image, what is the width of the actual snowflake? Express your answer as a decimal.

3. A community bike ride offers a short 5.7-mile ride for children and families. The short ride is followed by a long ride, $5\frac{2}{3}$ times as long as the short ride, for adults. If a woman bikes the short ride with her children, and then the long ride with her friends, how many miles does she ride altogether?

EUREKA MATH

Lesson 24:   Solve word problems using fraction and decimal multiplication.

105

4. Sal bought a house for \$78,524.60. Twelve years later he sold the house for $2\frac{3}{4}$ times as much. What was the sale price of the house?

5. In the fifth grade at Lenape Elementary School, there are $\frac{4}{5}$ as many students who do not wear glasses as those who do wear glasses. If there are 60 students who wear glasses, how many students are in the fifth grade?

6. At a factory, a mechanic earns \$17.25 an hour. The president of the company earns $6\frac{2}{3}$ times as much for each hour he works. The janitor at the same company earns $\frac{3}{5}$ as much as the mechanic. How much does the company pay for all three people employees' wages for one hour of work?

EUREKA MATH

Lesson 24: Solve word problems using fraction and decimal multiplication.

106

Name _____   Date _____

1. Draw a tape diagram and a number line to solve. You may draw the model that makes the most sense to you. Fill in the blanks that follow. Use the example to help you.

Example: $2 \div \frac{1}{3} =$ ___6___

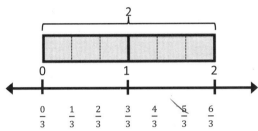   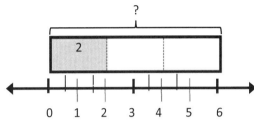

There are ___3___ thirds in 1 whole.

There are ___6___ thirds in 2 wholes.

If 2 is $\frac{1}{3}$, what is the whole? ___6___

a. $4 \div \frac{1}{2} =$ ___8___ / 4

There are __2__ halves in 1 whole.
There are __8__ halves in 4 wholes.

If 4 is $\frac{1}{2}$, what is the whole? ___8___

b. $2 \div \frac{1}{4} =$ ___8___

There are __4__ fourths in 1 whole.
There are __8__ fourths in 2 wholes.

If 2 is $\frac{1}{4}$, what is the whole? ___8___

c. $5 \div \frac{1}{3} =$ ___15___

There are __3__ thirds in 1 whole.
There are __15__ thirds in 5 wholes.

If 5 is $\frac{1}{3}$, what is the whole? ___15___

d. $3 \div \frac{1}{5} =$ ___15___

There are __5__ fifths in 1 whole.
There are __15__ fifths in 3 wholes.

If 3 is $\frac{1}{5}$, what is the whole? ___15___

EUREKA
MATH™

Lesson 25:   Divide a whole number by a unit fraction.

107

2. Divide. Then, multiply to check.

a. $5 \div \frac{1}{2} = 10$ $5 \times 2 = 10$ 5 $10 \times \frac{1}{2} = 5$	b. $3 \div \frac{1}{2} = 6$ $3 \times 2 = 6$ 3 $6 \times \frac{1}{2} = 3$	c. $4 \div \frac{1}{5} = 20$ $4 \times 5 = 20$ 4 $20 \times \frac{1}{5} = 4$	d. $1 \div \frac{1}{6} = 6$ $1 \times 6 = 6$ $6 \times \frac{1}{6} = 1$
e. $2 \div \frac{1}{8} = 16$ $2 \times 8 = 16$ $16 \times \frac{1}{8} = 2$	f. $7 \div \frac{1}{6} = 42$ $7 \times 6 = 42$ $42 \div \frac{1}{6} = 7$	g. $8 \div \frac{1}{3} = 24$ $8 \times 3 = 24$ $24 \div \frac{1}{3} = 8$	h. $9 \div \frac{1}{4} = 36$ $9 \times 4 = 36$ $36 \div \frac{1}{4} = 9$

3. For an art project, Mrs. Williams is dividing construction paper into fourths. How many fourths can she make from 5 pieces of construction paper?

$$5 \div \frac{1}{4} = 20 \text{ pieces}$$
$$5 \times 4 = 20$$

4. Use the chart below to answer the following questions.

Donnie's Diner Lunch Menu

Food	Serving Size
Hamburger	$\frac{1}{3}$ lb
Pickles	$\frac{1}{4}$ pickle
Potato chips	$\frac{1}{8}$ bag
Chocolate milk	$\frac{1}{2}$ cup

a. How many hamburgers can Donnie make with 6 pounds of hamburger meat?

$6 \times \frac{1}{3} = 18$ hamburgers

b. How many pickle servings can be made from a jar of 15 pickles?

$15 \times \frac{1}{4} = 60$ pickle servings

c. How many servings of chocolate milk can he serve from a gallon of milk?

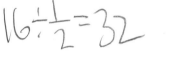 $16 \div \frac{1}{2} = 32$ 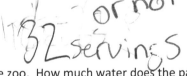 or not 32 servings

5. Three gallons of water fills $\frac{1}{4}$ of the elephant's pail at the zoo. How much water does the pail hold?

Name _____ Date _____

1. Draw a tape diagram and a number line to solve. Fill in the blanks that follow.

a. $3 \times 3 = 9$

$3 \div \frac{1}{3} =$ ___9___

There are __3__ thirds in 1 whole.

There are __9__ thirds in 3 wholes.

If 3 is $\frac{1}{3}$, what is the whole? ___9___

b. $3 \div \frac{1}{4} =$ ___12___

There are __4__ fourths in 1 whole.

There are __12__ fourths in __3__ wholes.

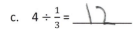

If 3 is $\frac{1}{4}$, what is the whole? ___12___

c. $4 \div \frac{1}{3} =$ ___12___

There are __3__ thirds in 1 whole.

There are __12__ thirds in __4__ wholes.

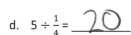

If 4 is $\frac{1}{3}$, what is the whole? ___12___

d. $5 \div \frac{1}{4} =$ ___20___

There are __4__ fourths in 1 whole.

There are __20__ fourths in __5__ wholes.

If 5 is $\frac{1}{4}$, what is the whole? ___20___

2. Divide. Then, multiply to check.

a. $2 \div \frac{1}{4} = 8$	b. $6 \div \frac{1}{2}$
$2 \times 4 = 8$	$6 \div \frac{1}{2} = 12$
$2 \div \frac{1}{4} = 8$	$6 \times 2 = 12$
$8 \times \frac{1}{4} = 2$	$12 \times \frac{1}{2} = 6$

c. $5 \div \frac{1}{4}$	d. $5 \div \frac{1}{8}$
$5 \div \frac{1}{4} = 20$	$5 \div \frac{1}{8} = 40$
$5 \times 4 = 20$	$5 \times 8 = 40$
$20 \times \frac{1}{4} = 5$	$40 \times \frac{1}{8} = 5$

e. $6 \div \frac{1}{3}$	f. $3 \div \frac{1}{6}$
$6 \div \frac{1}{3} = 18$	$3 \div \frac{1}{6} = 18$
$6 \times 3 = 18$	$6 \times 3 = 18$
$18 \times \frac{1}{3} = 6$	$18 \div \frac{1}{6} = 3$

g. $6 \div \frac{1}{5}$	h. $6 \div \frac{1}{10}$
$6 \div \frac{1}{5} = 30$	$6 \div \frac{1}{10} = 60$
$6 \times 5 = 30$	$6 \times 10 = 60$
$30 \times \frac{1}{5} = 6$	$60 \times \frac{1}{10} = 6$

Tape Diagram

3. A principal orders 8 sub sandwiches for a teachers' meeting. She cuts the subs into thirds and puts the mini-subs onto a tray. How many mini-subs are on the tray?

$8 \times 3 = 24$

24 mini-subs on the tray

$8 \div \frac{1}{3} = 24$

4. Some students prepare 3 different snacks. They make $\frac{1}{8}$ pound bags of nut mix, $\frac{1}{4}$ pound bags of cherries, and $\frac{1}{6}$ pound bags of dried fruit. If they buy 3 pounds of nut mix, 5 pounds of cherries, and 4 pounds of dried fruit, how many of each type of snack bag will they be able to make?

3 tape diagrams

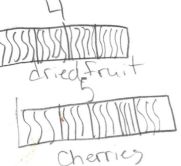

dried fruit

cherries

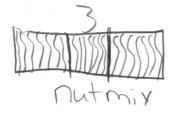

nutmix

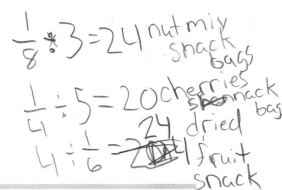

$\frac{1}{8} \div 3 = 24$ nut mix snack bags

$\frac{1}{4} \div 5 = 20$ cherries snack bags

$4 \div \frac{1}{6} = 24$ dried fruit snack bags.

Name _____   Date _____

1. Draw a model or tape diagram to solve. Use the thought bubble to show your thinking. Write your quotient in the blank. Use the example to help you.

Example: $\frac{1}{2} \div 3$

1 half ÷ 3

= 3 sixths ÷ 3

= 1 sixth

$\frac{1}{2} \div 3 = \frac{1}{6}$

a. $\frac{1}{3} \div 2 = \underline{\frac{1}{6}}$

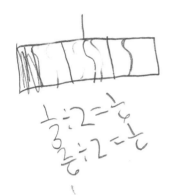

$\frac{1}{3} \div 2 = \frac{1}{6}$
$\frac{3}{6} \div 2 = \frac{1}{6}$

b. $\frac{1}{3} \div 4 = \underline{\frac{1}{2}}$

$\frac{1}{3} \div 4 = \frac{1}{12}$
$\frac{4}{12} \div 4 = \frac{1}{12}$

EUREKA MATH

Lesson 26:   Divide a unit fraction by a whole number.

c. $\frac{1}{4} \div 2 =$ ___$\frac{1}{8}$___

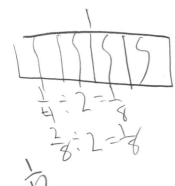

$\frac{1}{4} \div 2 = \frac{1}{8}$

$\frac{2}{8} \div 2 = \frac{1}{8}$

d. $\frac{1}{4} \div 3 =$ ___$\frac{1}{12}$___

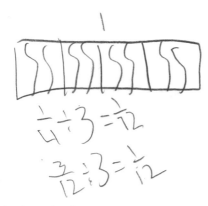

$\frac{1}{4} \div 3 = \frac{1}{12}$

$\frac{3}{12} \div 3 = \frac{1}{12}$

2.   Divide. Then, multiply to check.

a. $\frac{1}{2} \div 7$	b. $\frac{1}{3} \div 6 = \frac{1}{18}$	c. $\frac{1}{4} \div 5 = \frac{1}{20}$	d. $\frac{1}{5} \div 4 = \frac{1}{20}$
$\frac{1}{2} \div 7 = \frac{1}{14}$ $\frac{7}{14} \div 7 = \frac{1}{14}$ $\frac{1}{14} \times 7 = \frac{1}{2}$	$\frac{6}{18} \div 6 = \frac{1}{18}$ $\frac{1}{8} \times 6 = \frac{1}{3}$	$\frac{5}{20} \div 5 = \frac{1}{20}$ $\frac{1}{20} \times 5 = \frac{1}{4}$	$\frac{4}{20} \div 4 = \frac{1}{20}$ $\frac{1}{20} \times 4 = \frac{1}{5}$
e. $\frac{1}{5} \div 2 = \frac{1}{10}$	f. $\frac{1}{6} \div 3 = \frac{1}{18}$	g. $\frac{1}{8} \div 2 = \frac{1}{16}$	h. $\frac{1}{10} \div 10 = \frac{1}{100}$
$\frac{2}{10} \div 2 = \frac{1}{10}$ $\frac{1}{10} \times 2 = \frac{1}{5}$	$\frac{3}{18} \div 3 = \frac{1}{18}$ $\frac{1}{18} \times 3 = \frac{1}{6}$	$\frac{2}{16} \div 2 = \frac{1}{16}$ $\frac{1}{16} \times 2 = \frac{1}{8}$	$\frac{10}{100} \div 10 = \frac{1}{100}$ $\frac{1}{100} \times 10 = \frac{1}{10}$

3. Tasha eats half her snack and gives the other half to her two best friends for them to share equally. What portion of the whole snack does each friend get? Draw a picture to support your response.

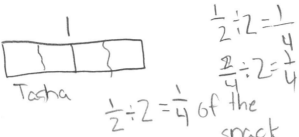

4. Mrs. Appler used $\frac{1}{2}$ gallon of olive oil to make 8 identical batches of salad dressing.

    a. How many gallons of olive oil did she use in each batch of salad dressing?

    b. How many cups of olive oil did she use in each batch of salad dressing?

5. Mariano delivers newspapers. He always puts $\frac{3}{4}$ of his weekly earnings in his savings account, and then divides the rest equally into 3 piggy banks for spending at the snack shop, the arcade, and the subway.

    a. What fraction of his earnings does Mariano put into each piggy bank?

    b. If Mariano adds $2.40 to each piggy bank every week, how much does Mariano earn per week delivering papers?

EUREKA
MATH™   Lesson 26:   Divide a unit fraction by a whole number.

115

Name _____ Date _____

1. Solve and support your answer with a model or tape diagram. Write your quotient in the blank.

a. $\frac{1}{2} \div 4 = \underline{\frac{1}{8}}$

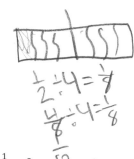

$\frac{1}{2} \div 4 = \frac{1}{8}$

$\frac{4}{8} \div 4 = \frac{1}{8}$

b. $\frac{1}{3} \div 6 = \underline{\frac{1}{18}}$

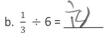

$\frac{1}{3} \div 6 = \frac{1}{18}$

$\frac{6}{18} \div 6 = \frac{1}{3}$

c. $\frac{1}{4} \div 3 = \underline{\frac{1}{12}}$

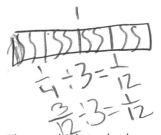

$\frac{1}{4} \div 3 = \frac{1}{12}$

$\frac{3}{12} \div 3 = \frac{1}{12}$

d. $\frac{1}{5} \div 2 = \underline{\frac{1}{10}}$

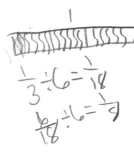

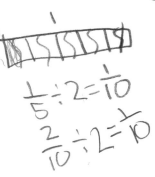

$\frac{1}{5} \div 2 = \frac{1}{10}$

$\frac{2}{10} \div 2 = \frac{1}{10}$

2. Divide. Then, multiply to check.

a. $\frac{1}{2} \div 10 = \frac{1}{20}$	b. $\frac{1}{4} \div 10 = \frac{1}{40}$	c. $\frac{1}{3} \div 5 = \frac{1}{15}$	d. $\frac{1}{5} \div 3 = \frac{1}{15}$
$\frac{10}{20} \div 10 = \frac{1}{20}$ $\frac{1}{20} \times 10 = \frac{1}{2}$	$\frac{10}{40} \div 10 = \frac{1}{40}$ $\frac{1}{40} \times 10 = \frac{1}{4}$	$\frac{5}{15} \div 5 = \frac{1}{15}$ $\frac{1}{15} \times 5 = \frac{1}{3}$	$\frac{3}{15} \div 3 = \frac{1}{15}$ $\frac{1}{15} \times 3 = \frac{1}{5}$
e. $\frac{1}{8} \div 4 = \frac{1}{32}$	f. $\frac{1}{7} \div 3 = \frac{1}{21}$	g. $\frac{1}{10} \div 5 = \frac{1}{50}$	h. $\frac{1}{5} \div 20 = \frac{1}{100}$
$\frac{4}{32} \div 4 = \frac{1}{32}$ $\frac{1}{32} \times 4 = \frac{1}{8}$	$\frac{3}{21} \div 3 = \frac{1}{21}$ $\frac{1}{21} \times 3 = \frac{1}{7}$	$\frac{5}{50} \div 5 = \frac{1}{50}$ $\frac{1}{50} \times 5 = \frac{1}{10}$	$\frac{20}{100} \div 20 = \frac{1}{100}$ $\frac{1}{100} \times 20 = \frac{1}{5}$

3. Teams of four are competing in a quarter-mile relay race. Each runner must run the same exact distance. What is the distance each teammate runs?

$$\frac{1}{4} \div 4 = \frac{1}{16} \text{ of a quarter mile}$$

4. Solomon has read $\frac{1}{3}$ of his book. He finishes the book by reading the same amount each night for 5 nights.

   a. What fraction of the book does he read each of the 5 nights?

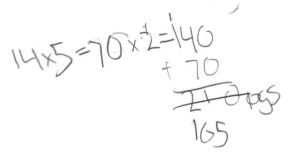

   $$\frac{1}{3} \div 5 = \frac{1}{15}$$

   $$\frac{5}{15} \div 5 = \frac{1}{5}$$

   b. If he reads 14 pages on each of the 5 nights, how long is the book?

   $$14 \times 5 = 70 \times 2 = 140$$
   $$+ 70$$
   ~~170 pgs~~
   $$165$$

EUREKA MATH   **Lesson 26:**   Divide a unit fraction by a whole number.

117

Name _____ Date _____

1. Mrs. Silverstein bought 3 mini cakes for a birthday party. She cuts each cake into quarters and plans to serve each guest 1 quarter of a cake. How many guests can she serve with all her cakes? Draw a picture to support your response.

$3 \div \frac{1}{4} = 12$ guests

2. Mr. Pham has $\frac{1}{4}$ pan of lasagna left in the refrigerator. He wants to cut the lasagna into equal slices so he can have it for dinner for 3 nights. How much lasagna will he eat each night? Draw a picture to support your response.

$$\frac{1}{4} \div 3 = \frac{1}{12}$$

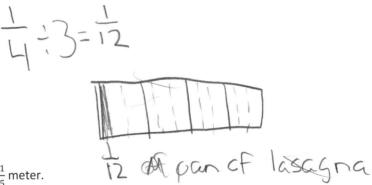

$\frac{1}{12}$ of a pan of lasagna

3. The perimeter of a square is $\frac{1}{5}$ meter.
   a. Find the length of each side in meters. Draw a picture to support your response.

$$\frac{1}{5} \div 4 = \frac{1}{20} \text{ meter}$$

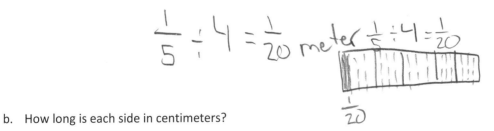

$\frac{1}{5} \div 4 = \frac{1}{20}$

$\frac{1}{20}$

   b. How long is each side in centimeters?

4. A pallet holding 5 identical crates weighs $\frac{1}{4}$ ton.

   a. How many tons does each crate weigh? Draw a picture to support your response.

   $$\frac{1}{4} \div 5$$

   $$\frac{1}{4} \div 5 = \frac{1}{20} \text{ tons}$$

   b. How many pounds does each crate weigh?

5. Faye has 5 pieces of ribbon, each 1 yard long. She cuts each ribbon into sixths.

   a. How many sixths will she have after cutting all the ribbons?

   $$5 \div \frac{1}{6} = 30 \text{ sixths}$$

   b. How long will each of the sixths be in inches?

6. A glass pitcher is filled with water. $\frac{1}{8}$ of the water is poured equally into 2 glasses.

   a. What fraction of the water is in each glass?

$$\frac{1}{8} \div 2 = \frac{1}{16} \text{ of the water}$$

   b. If each glass has 3 fluid ounces of water in it, how many fluid ounces of water were in the full pitcher?

$$6 \text{ oz} \div \frac{1}{8} = 48 \text{ ounces}$$

   c. If $\frac{1}{4}$ of the remaining water is poured out of the pitcher to water a plant, how many cups of water are left in the pitcher?

Name _____ Date _____

1. Kelvin ordered four pizzas for a birthday party. The pizzas were cut in eighths. How many slices were there? Draw a picture to support your response.

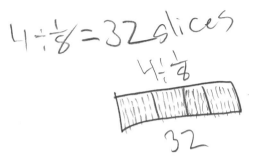

$$4 \div \tfrac{1}{8} = 32 \text{ slices}$$

2. Virgil has $\frac{1}{6}$ of a birthday cake left over. He wants to share the leftover cake with 3 friends. What fraction of the original cake will each of the 4 people receive? Draw a picture to support your response.

$$\tfrac{1}{6} \div 4 = \tfrac{1}{24}$$

3. A pitcher of water contains $\frac{1}{4}$ liters of water. The water is poured equally into 5 glasses.

   a. How many liters of water are in each glass? Draw a picture to support your response.

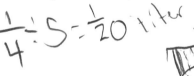

$$\tfrac{1}{4} \div 5 = \tfrac{1}{20} \text{ liter}$$

   b. Write the amount of water in each glass in milliliters.

4. Drew has 4 pieces of rope 1 meter long each. He cuts each rope into fifths.
   a. How many fifths will he have after cutting all the ropes?

   $$4 \div \frac{1}{5} = 20 \text{ fifths}$$

   b. How long will each of the fifths be in centimeters?

5. A container is filled with blueberries. $\frac{1}{6}$ of the blueberries is poured equally into two bowls.
   a. What fraction of the blueberries is in each bowl?

   $$\frac{1}{6} \div 2 = \frac{1}{12} \text{ of the blueberries}$$

   b. If each bowl has 6 ounces of blueberries in it, how many ounces of blueberries were in the full container?

   $$12 \div \frac{1}{6} = 72 \text{ ounces}$$

   Extra credit

   c. If $\frac{1}{5}$ of the remaining blueberries are used to make muffins, how many pounds of blueberries are left in the container?

   $$16 oz = 1 lb$$

   $$72 - 12 = 60 \qquad 60 - 12 = 48 \text{ ounces} \quad \frac{16}{\times 3} = 48$$

   3 pounds of blueberries

Name _____   Date _____

1. Create and solve a division story problem about 5 meters of rope that is modeled by the tape diagram below.

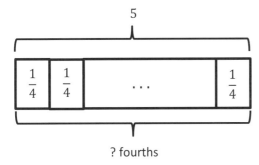

? fourths

2. Create and solve a story problem about $\frac{1}{4}$ pound of almonds that is modeled by the tape diagram below.

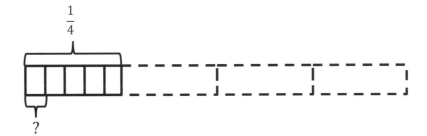

EUREKA MATH™

Lesson 28:   Write equations and word problems corresponding to tape and number line diagrams.

125

3. Draw a tape diagram and create a word problem for the following expressions, and then solve.

    a. $2 \div \frac{1}{3}$

    b. $\frac{1}{3} \div 4$

    c. $\frac{1}{4} \div 3$

    d. $3 \div \frac{1}{5}$

EUREKA MATH™ | **Lesson 28:**   Write equations and word problems corresponding to tape and number line diagrams.

126

Name _____   Date _____

1. Create and solve a division story problem about 7 feet of rope that is modeled by the tape diagram below.

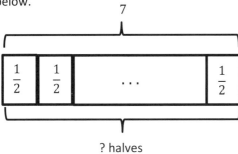

2. Create and solve a story problem about $\frac{1}{3}$ pound of flour that is modeled by the tape diagram below.

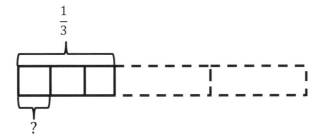

EUREKA MATH   Lesson 28:   Write equations and word problems corresponding to tape and number line diagrams.

127

3. Draw a tape diagram and create a word problem for the following expressions. Then, solve and check.

   a. $2 \div \frac{1}{4}$

   b. $\frac{1}{4} \div 2$

   c. $\frac{1}{3} \div 5$

   d. $3 \div \frac{1}{10}$

EUREKA MATH

Lesson 28: Write equations and word problems corresponding to tape and number line diagrams.

128

Name _____ Date _____

1. Divide. Rewrite each expression as a division sentence with a fraction divisor, and fill in the blanks. The first one is done for you.

Example: $2 \div 0.1 = 2 \div \frac{1}{10} = 20$

There are __10__ tenths in 1 whole.

There are __20__ tenths in 2 wholes.

a. $5 \div 0.1 =$

There are _____ tenths in 1 whole.

There are _____ tenths in 5 wholes.

b. $8 \div 0.1 =$

There are _____ tenths in 1 whole.

There are _____ tenths in 8 wholes.

c. $5.2 \div 0.1 =$

There are _____ tenths in 5 wholes.

There are _____ tenths in 2 tenths.

There are _____ tenths in 5.2

d. $8.7 \div 0.1 =$

There are _____ tenths in 8 wholes.

There are _____ tenths in 7 tenths.

There are _____ tenths in 8.7

e. $5 \div 0.01 =$

There are _____ hundredths in 1 whole.

There are _____ hundredths in 5 wholes.

f. $8 \div 0.01 =$

There are _____ hundredths in 1 whole.

There are _____ hundredths in 8 wholes.

g. $5.2 \div 0.01 =$

There are _____ hundredths in 5 wholes.

There are _____ hundredths in 2 tenths.

There are _____ hundredths in 5.2

h. $8.7 \div 0.01 =$

There are _____ hundredths in 8 wholes.

There are _____ hundredths in 7 tenths.

There are _____ hundredths in 8.7

Lesson 29: Connect division by a unit fraction to division by 1 tenth and 1 hundredth.

2. Divide.

a. 6 ÷ 0.1	b. 18 ÷ 0.1	c. 6 ÷ 0.01
d. 1.7 ÷ 0.1	e. 31 ÷ 0.01	f. 11 ÷ 0.01
g. 125 ÷ 0.1	h. 3.74 ÷ 0.01	i. 12.5 ÷ 0.01

3. Yung bought $4.60 worth of bubble gum. Each piece of gum cost $0.10. How many pieces of bubble gum did Yung buy?

4. Cheryl solved a problem: 84 ÷ 0.01 = 8,400.
Jane said, "Your answer is wrong because when you divide, the quotient is always smaller than the whole amount you start with, for example, 6 ÷ 2 = 3 and 100 ÷ 4 = 25." Who is correct? Explain your thinking.

5. The U.S. Mint sells 2 ounces of American Eagle gold coins to a collector. Each coin weighs one-tenth of an ounce. How many gold coins were sold to the collector?

EUREKA MATH™

Lesson 29:   Connect division by a unit fraction to division by 1 tenth and
1 hundredth.

130

Name _____ Date _____

1. Divide. Rewrite each expression as a division sentence with a fraction divisor, and fill in the blanks. The first one is done for you.

Example: $4 \div 0.1 = 4 \div \frac{1}{10} = 40$

There are __10__ tenths in 1 whole.

There are __40__ tenths in 4 wholes.

a. $9 \div 0.1 =$

There are _____ tenths in 1 whole.

There are _____ tenths in 9 wholes.

b. $6 \div 0.1 =$

There are _____ tenths in 1 whole.

There are _____ tenths in 6 wholes.

c. $3.6 \div 0.1 =$

There are _____ tenths in 3 wholes.

There are _____ tenths in 6 tenths.

There are _____ tenths in 3.6.

d. $12.8 \div 0.1 =$

There are _____ tenths in 12 wholes.

There are _____ tenths in 8 tenths.

There are _____ tenths in 12.8.

e. $3 \div 0.01 =$

There are _____ hundredths in 1 whole.

There are _____ hundredths in 3 wholes.

f. $7 \div 0.01 =$

There are _____ hundredths in 1 whole.

There are _____ hundredths in 7 wholes.

g. $4.7 \div 0.01 =$

There are _____ hundredths in 4 wholes.

There are _____ hundredths in 7 tenths.

There are _____ hundredths in 4.7.

h. $11.3 \div 0.01 =$

There are _____ hundredths in 11 wholes.

There are _____ hundredths in 3 tenths.

There are _____ hundredths in 11.3.

EUREKA MATH™

Lesson 29: Connect division by a unit fraction to division by 1 tenth and 1 hundredth.

131

2. Divide.

a. $2 \div 0.1$	b. $23 \div 0.1$	c. $5 \div 0.01$
d. $7.2 \div 0.1$	e. $51 \div 0.01$	f. $31 \div 0.1$
g. $231 \div 0.1$	h. $4.37 \div 0.01$	i. $24.5 \div 0.01$

3. Giovanna is charged $0.01 for each text message she sends. Last month, her cell phone bill included a $12.60 charge for text messages. How many text messages did Giovanna send?

4. Geraldine solved a problem: $68.5 \div 0.01 = 6{,}850$.
   Ralph said, "This is wrong because a quotient can't be greater than the whole you start with. For example, $8 \div 2 = 4$ and $250 \div 5 = 50$." Who is correct? Explain your thinking.

5. The price for an ounce of gold on September 23, 2013, was $1,326.40. A group of 10 friends decide to equally share the cost of 1 ounce of gold. How much money will each friend pay?

EUREKA
MATH™

Lesson 29: Connect division by a unit fraction to division by 1 tenth and
1 hundredth.

132

Name _____   Date _____

1. Rewrite the division expression as a fraction and divide. The first two have been started for you.

a. $2.7 \div 0.3 = \dfrac{2.7}{0.3}$ $= \dfrac{2.7 \times 10}{0.3 \times 10}$ $= \dfrac{27}{3}$ $= 9$	b. $2.7 \div 0.03 = \dfrac{2.7}{0.03}$ $= \dfrac{2.7 \times 100}{0.03 \times 100}$ $= \dfrac{270}{3}$ $=$
c. $3.5 \div 0.5 =$	d. $3.5 \div 0.05 =$
e. $4.2 \div 0.7 =$	f. $0.42 \div 0.07 =$

EUREKA MATH

Lesson 30: Divide decimal dividends by non-unit decimal divisors.

133

g. $10.8 \div 0.9 =$	h. $1.08 \div 0.09 =$
i. $3.6 \div 1.2 =$	j. $0.36 \div 0.12 =$
k. $17.5 \div 2.5 =$	l. $1.75 \div 0.25 =$

2. $15 \div 3 = 5$. Explain why it is true that $1.5 \div 0.3$ and $0.15 \div 0.03$ have the same quotient.

3. Mr. Volok buys 2.4 kg of sugar for his bakery.

    a. If he pours 0.2 kg of sugar into separate bags, how many bags of sugar can he make?

    b. If he pours 0.4 kg of sugar into separate bags, how many bags of sugar can he make?

4. Two wires, one 17.4 meters long and one 7.5 meters long, were cut into pieces 0.3 meters long. How many such pieces can be made from both wires?

5. Mr. Smith has 15.6 pounds of oranges to pack for shipment. He can ship 2.4 pounds of oranges in a large box and 1.2 pounds in a small box. If he ships 5 large boxes, what is the minimum number of small boxes required to ship the rest of the oranges?

Lesson 30:   Divide decimal dividends by non-unit decimal divisors.

135

Name _____   Date _____

1. Rewrite the division expression as a fraction and divide. The first two have been started for you.

a. $2.4 \div 0.8 = \dfrac{2.4}{0.8}$ $= \dfrac{2.4 \times 10}{0.8 \times 10}$ $= \dfrac{24}{8}$ $=$	b. $2.4 \div 0.08 = \dfrac{2.4}{0.08}$ $= \dfrac{2.4 \times 100}{0.08 \times 100}$ $= \dfrac{240}{8}$ $=$
c. $4.8 \div 0.6 =$	d. $0.48 \div 0.06 =$
e. $8.4 \div 0.7 =$	f. $0.84 \div 0.07 =$

Lesson 30:   Divide decimal dividends by non-unit decimal divisors.

g. $4.5 \div 1.5 =$	h. $0.45 \div 0.15 =$
i. $14.4 \div 1.2 =$	j. $1.44 \div 0.12 =$

2. Leann says $18 \div 6 = 3$, so $1.8 \div 0.6 = 0.3$ and $0.18 \div 0.06 = 0.03$. Is Leann correct? Explain how to solve these division problems.

3. Denise is making bean bags. She has 6.4 pounds of beans.

   a. If she makes each bean bag 0.8 pounds, how many bean bags will she be able to make?

   b. If she decides instead to make mini bean bags that are half as heavy, how many can she make?

EUREKA MATH™   Lesson 30:   Divide decimal dividends by non-unit decimal divisors.

137

4. A restaurant's small salt shakers contain 0.6 ounces of salt. Its large shakers hold twice as much. The shakers are filled from a container that has 18.6 ounces of salt. If 8 large shakers are filled, how many small shakers can be filled with the remaining salt?

Lesson 30:   Divide decimal dividends by non-unit decimal divisors.

138

Name _____ Date _____

1. Estimate, and then divide. An example has been done for you.

$78.4 \div 0.7 \approx 770 \div 7 = 110$

$= \dfrac{78.4}{0.7}$

$= \dfrac{78.4 \times 10}{0.7 \times 10}$

$= \dfrac{784}{7}$

$= 112$

```
      1 1 2
7 | 7 8 4
    -7
     8
    -7
     1 4
    -1 4
      0
```

a. $53.2 \div 0.4 \approx$

b. $1.52 \div 0.8 \approx$

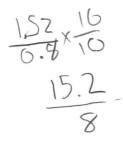

2. Estimate, and then divide. The first one has been done for you.

$7.32 \div 0.06 \approx 720 \div 6 = 120$

$= \dfrac{7.32}{0.06}$

$= \dfrac{7.32 \times 100}{0.06 \times 100}$

$= \dfrac{732}{6}$

$= 122$

```
      1 2 2
6 | 7 3 2
    -6
     1 3
    -1 2
      1 2
     -1 2
       0
```

a. $9.42 \div 0.03 \approx$

b. $39.36 \div 0.96 \approx$

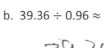

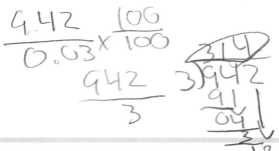

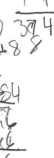

Lesson 31:   Divide decimal dividends by non-unit decimal divisors.

139

3. Solve using the standard algorithm. Use the thought bubble to show your thinking as you rename the divisor as a whole number.

a. 46.2 ÷ 0.3 = _____

$3 \overline{)4\ 6\ 2}$

$= \frac{6.2}{0.3} = 154$

b. 3.16 ÷ 0.04 = _____

c. 2.31 ÷ 0.3 = 7.7

$3\overline{)23.1}$
21
21
21

$\frac{2.31}{0.3} \times \frac{10}{10}$

$\frac{23.1}{3}$

d. 15.6 ÷ 0.24 = 65

$\frac{15.6}{0.24} \times \frac{100}{100}$

$\frac{1560}{24}$

4. The total distance of a race is 18.9 km.
   a. If volunteers set up a water station every 0.7 km, including one at the finish line, how many stations will they have?

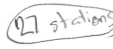

27 stations

$\frac{18.9}{0.7} \times \frac{10}{10}$

$\frac{189}{7}$

   b. If volunteers set up a first aid station every 0.9 km, including one at the finish line, how many stations will they have?

5. In a laboratory, a technician combines a salt solution contained in 27 test tubes. Each test tube contains 0.06 liter of the solution. If he divides the total amount into test tubes that hold 0.3 liter each, how many test tubes will he need?

EUREKA MATH

Lesson 31:   Divide decimal dividends by non-unit decimal divisors.

140

Name _____   Date _____

1. Estimate, and then divide. An example has been done for you.

$78.4 \div 0.7 \approx 770 \div 7 = 110$

$$= \frac{78.4}{0.7}$$

$$= \frac{78.4 \times 10}{0.7 \times 10}$$

$$= \frac{784}{7}$$

$$= 112$$

```
        1 1 2
   7 | 7 8 4
       -7
        8
       -7
        1 4
       -1 4
        0
```

a. $61.6 \div 0.8 \approx$

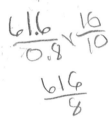

b. $5.74 \div 0.7 \approx$

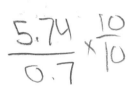

2. Estimate, and then divide. An example has been done for you.

$7.32 \div 0.06 \approx 720 \div 6 = 120$

$$= \frac{7.32}{0.06}$$

$$= \frac{7.32 \times 100}{0.06 \times 100}$$

$$= \frac{732}{6}$$

$$= 122$$

```
        1 2 2
   6 | 7 3 2
       -6
        1 3
       -1 2
        1 2
       -1 2
        0
```

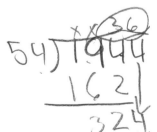

a. $4.74 \div 0.06 \approx$

b. $19.44 \div 0.54 \approx$

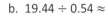
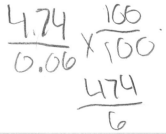

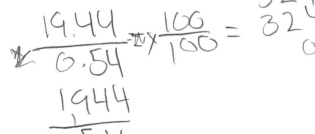

Lesson 31:   Divide decimal dividends by non-unit decimal divisors.

141

3. Solve using the standard algorithm. Use the thought bubble to show your thinking as you rename the divisor as a whole number.

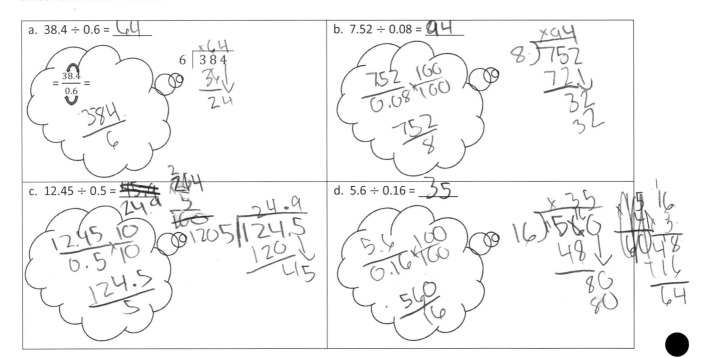

a. $38.4 \div 0.6 = 64$	b. $7.52 \div 0.08 = 94$
c. $12.45 \div 0.5 = 24.9$	d. $5.6 \div 0.16 = 35$

4. Lucia is making a 21.6 centimeter beaded string to hang in the window. She decides to put a green bead every 0.4 centimeters and a purple bead every 0.6 centimeters. How many green beads and how many purple beads will she need?

$$\frac{21.6}{0.6} \times \frac{10}{10} \qquad \frac{21.6}{0.4} \times \frac{10}{10} = \frac{216}{4}$$

$$\frac{216}{6}$$

54 green beads
36 purple beads

5. A group of 14 friends collects 0.7 pound of blueberries and decides to make blueberry muffins. They put 0.05 pound of berries in each muffin. How many muffins can they make if they use all the blueberries they collected?

$$\frac{0.7}{0.05} \times \frac{100}{100} = 0.7 \div 0.05$$

$$\frac{70}{5}$$

14 muffins

Name _____ Date _____

1. Circle the expression equivalent to *the sum of 3 and 2 divided by* $\frac{1}{3}$.

   $\frac{3+2}{3}$ $3 + (2 \div \frac{1}{3})$ $(3 + 2) \div \frac{1}{3}$ $\frac{1}{3} \div (3 + 2)$

2. Circle the expression(s) equivalent to *28 divided by the difference between* $\frac{4}{5}$ *and* $\frac{7}{10}$.

   $28 \div \left(\frac{4}{5} - \frac{7}{10}\right)$ $\frac{28}{\frac{4}{5} - \frac{7}{10}}$ $\left(\frac{4}{5} - \frac{7}{10}\right) \div 28$ $28 \div \left(\frac{7}{10} - \frac{4}{5}\right)$

3. Fill in the chart by writing an equivalent numerical expression.

a.	Half as much as the difference between $2\frac{1}{4}$ and $\frac{3}{8}$.	
b.	The difference between $2\frac{1}{4}$ and $\frac{3}{8}$ divided by 4.	
c.	A third of the sum of $\frac{7}{8}$ and 22 tenths.	
d.	Add 2.2 and $\frac{7}{8}$, and then triple the sum.	

4. Compare expressions 3(a) and 3(b). Without evaluating, identify the expression that is greater. Explain how you know.

EUREKA MATH

Lesson 32: Interpret and evaluate numerical expressions including the language of scaling and fraction division.

143

5. Fill in the chart by writing an equivalent expression in word form.

a.		$\frac{3}{4} \times (1.75 + \frac{3}{5})$
b.		$\frac{7}{9} - (\frac{1}{8} \times 0.2)$
c.		$(1.75 + \frac{3}{5}) \times \frac{4}{3}$
d.		$2 \div (\frac{1}{2} \times \frac{4}{5})$

6. Compare the expressions in 5(a) and 5(c). Without evaluating, identify the expression that is less. Explain how you know.

7. Evaluate the following expressions.

a. $(9 - 5) \div \frac{1}{3}$

b. $\frac{5}{3} \times (2 \times \frac{1}{4})$

c. $\frac{1}{3} \div (1 \div \frac{1}{4})$

d. $\frac{1}{2} \times \frac{3}{5} \times \frac{5}{3}$

e. Half as much as $(\frac{3}{4} \times 0.2)$

f. 3 times as much as the quotient of 2.4 and 0.6

8. Choose an expression below that matches the story problem, and write it in the blank.

$$\frac{2}{3} \times (20 - 5) \qquad (\frac{2}{3} \times 20) - (\frac{2}{3} \times 5) \qquad \frac{2}{3} \times 20 - 5 \qquad (20 - \frac{2}{3}) - 5$$

a. Farmer Green picked 20 carrots. He cooked $\frac{2}{3}$ of them, and then gave 5 to his rabbits. Write the expression that tells how many carrots he had left.

Expression: _____

b. Farmer Green picked 20 carrots. He cooked 5 of them, and then gave $\frac{2}{3}$ to his rabbits. Write the expression that tells how many carrots the rabbits will get.

Expression: _____

EUREKA
MATH™

Lesson 32:   Interpret and evaluate numerical expressions including the language of scaling and fraction division.

145

Name _____   Date _____

1. Circle the expression equivalent to *the difference between 7 and 4, divided by a fifth*.

$$7 + (4 \div \tfrac{1}{5}) \qquad \qquad \frac{7-4}{5} \qquad \qquad (7-4) \div \tfrac{1}{5} \qquad \qquad \tfrac{1}{5} \div (7-4)$$

2. Circle the expression(s) equivalent to *42 divided by the sum of $\frac{2}{3}$ and $\frac{3}{4}$*.

$$(\tfrac{2}{3} + \tfrac{3}{4}) \div 42 \qquad (42 \div \tfrac{2}{3}) + \tfrac{3}{4} \qquad 42 \div (\tfrac{2}{3} + \tfrac{3}{4}) \qquad \frac{42}{\tfrac{2}{3} + \tfrac{3}{4}}$$

3. Fill in the chart by writing the equivalent numerical expression or expression in word form.

	Expression in word form	Numerical expression
a.	A fourth as much as the sum of $3\frac{1}{8}$ and 4.5	
b.		$(3\frac{1}{8} + 4.5) \div 5$
c.	Multiply $\frac{3}{5}$ by 5.8, then halve the product	
d.		$\frac{1}{6} \times (4.8 - \frac{1}{2})$
e.		$8 - (\frac{1}{2} \div 9)$

4. Compare the expressions in 3(a) and 3(b). Without evaluating, identify the expression that is greater. Explain how you know.

EUREKA
MATH™

Lesson 32:   Interpret and evaluate numerical expressions including the language of scaling and fraction division.

146

5. Evaluate the following expressions.

   a. $(11 - 6) \div \frac{1}{6}$

   b. $\frac{9}{5} \times (4 \times \frac{1}{6})$

   c. $\frac{1}{10} \div (5 \div \frac{1}{2})$

   d. $\frac{3}{4} \times \frac{2}{5} \times \frac{4}{3}$

   e. 50 divided by the difference between $\frac{3}{4}$ and $\frac{5}{8}$

6. Lee is sending out 32 birthday party invitations. She gives 5 invitations to her mom to give to family members. Lee mails a third of the rest, and then she takes a break to walk her dog.

   a. Write a numerical expression to describe how many invitations Lee has already mailed.

   b. Which expression matches how many invitations still need to be sent out?

   $32 - 5 - \frac{1}{3}(32 - 5)$ $\frac{2}{3} \times 32 - 5$ $(32 - 5) \div \frac{1}{3}$ $\frac{1}{3} \times (32 - 5)$

EUREKA MATH™

Lesson 32: Interpret and evaluate numerical expressions including the language of scaling and fraction division.

147

Name _____   Date _____

1. Ms. Hayes has $\frac{1}{2}$ liter of juice. She distributes it equally to 6 students in her tutoring group.

   a. How many liters of juice does each student get?

   b. How many more liters of juice will Ms. Hayes need if she wants to give each of the 24 students in her class the same amount of juice found in Part (a)?

2. Lucia has 3.5 hours left in her workday as a car mechanic. Lucia needs $\frac{1}{2}$ of an hour to complete one oil change.
   a. How many oil changes can Lucia complete during the rest of her workday?

   b. Lucia can complete two car inspections in the same amount of time it takes her to complete one oil change. How long does it take her to complete one car inspection?

   c. How many inspections can she complete in the rest of her workday?

 Lesson 33: Create story contexts for numerical expressions and tape diagrams, and solve word problems.

149

3. Carlo buys $14.40 worth of grapefruit. Each grapefruit costs $0.80.

    a. How many grapefruist does Carlo buy?

    b. At the same store, Kahri spends one-third as much money on grapefruits as Carlo. How many grapefruits does she buy?

4. Studies show that a typical giant hummingbird can flap its wings once in 0.08 of a second.

    a. While flying for 7.2 seconds, how many times will a typical giant hummingbird flap its wings?

    b. A ruby-throated hummingbird can flap its wings 4 times faster than a giant hummingbird. How many times will a ruby-throated hummingbird flap its wings in the same amount of time?

5. Create a story context for the following expression.

$$\frac{1}{3} \times (\$20 - \$3.20)$$

6. Create a story context about painting a wall for the following tape diagram.

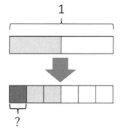

Lesson 33:   Create story contexts for numerical expressions and tape diagrams,
and solve word problems.

151

Name _____   Date _____

1. Chase volunteers at an animal shelter after school, feeding and playing with the cats.

   a. If he can make 5 servings of cat food from a third of a kilogram of food, how much does one serving weigh?

   b. If Chase wants to give this same serving size to each of 20 cats, how many kilograms of food will he need?

2. Anouk has 4.75 pounds of meat. She uses a quarter pound of meat to make one hamburger.

   a. How many hamburgers can Anouk make with the meat she has?

   b. Sometimes Anouk makes sliders. Each slider is half as much meat as is used for a regular hamburger. How many sliders could Anouk make with the 4.75 pounds?

EUREKA
MATH™

Lesson 33:   Create story contexts for numerical expressions and tape diagrams,
and solve word problems.

152

3. Ms. Geronimo has a $10 gift certificate to her local bakery.

   a. If she buys a slice of pie for $2.20 and uses the rest of the gift certificate to buy chocolate macaroons that cost $0.60 each, how many macaroons can Ms. Geronimo buy?

   b. If she changes her mind and instead buys a loaf of bread for $4.60 and uses the rest to buy cookies that cost $1\frac{1}{2}$ times as much as the macaroons, how many cookies can she buy?

4. Create a story context for the following expressions.

   a. $(5\frac{1}{4} - 2\frac{1}{8}) \div 4$

   b. $4 \times (\frac{4.8}{0.8})$

5. Create a story context for the following tape diagram.

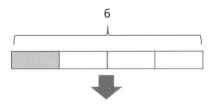

EUREKA
MATH™

Lesson 33: Create story contexts for numerical expressions and tape diagrams, and solve word problems.

153